Synthesis Lectures on Mathematics & Statistics

Series Editor

Steven G. Krantz, Department of Mathematics, Washington University, Saint Louis, MO, USA

This series includes titles in applied mathematics and statistics for cross-disciplinary STEM professionals, educators, researchers, and students. The series focuses on new and traditional techniques to develop mathematical knowledge and skills, an understanding of core mathematical reasoning, and the ability to utilize data in specific applications.

Daniel Arrigo

Analytical Methods for Solving Nonlinear Partial Differential Equations

Second Edition

 Springer

Daniel Arrigo
University of Central Arkansas
Conway, AR, USA

ISSN 1938-1743   ISSN 1938-1751   (electronic)
Synthesis Lectures on Mathematics & Statistics
ISBN 978-3-031-17071-3   ISBN 978-3-031-17069-0   (eBook)
https://doi.org/10.1007/978-3-031-17069-0

Preface to Second Edition

The second edition primarily mirrors the first edition. Several typos have been corrected and new material has been added. In particular, "Generating Contact transformations" and "Parametric Legendre transformation" both seen in Chap. 4. Several addition exercises and references have been added to supplement these sections.

Conway, USA
September 2022

Daniel Arrigo

Preface to First Edition

This is somewhat of an introductory book about obtaining exact solutions to nonlinear partial differential equations (NLPDEs). This book is based, in part, on my lecture notes from a graduate course I have taught at the University of Central Arkansas (UCA) for over a decade.

In Chap. 1, I list several nonlinear PDEs (systems of PDEs) that appear in science and engineering and provide a springboard into the subject matter. As our world is essentially nonlinear, it is not surprising that the equations that govern most physical phenomena are nonlinear. As Nobel Laureate Werner Heisenberg once said, "The progress of physics will to a large extent depend on the progress of nonlinear mathematics, of methods to solve nonlinear equations."

In Chap. 2, I introduce compatibility. I start with the method of characteristics, a topic from a standard course in PDEs. I then move to Charpit's method, a method that seeks compatibility between two first order PDEs. From there, I consider the compatibility between first and second order PDEs, particularly, a nonlinear diffusion equation, Burgers' equation and a linear diffusion equation with a nonlinear source term. I then extend the method to the nonlinear diffusion equations in $(2 + 1)$ dimensions and a system of nonlinear PDEs where we consider the Cubic Schrödinger equation.

In Chap. 3, I introduce the idea of differential substitutions. The classic example is the Hopf-Cole transformation and Burgers' equation; it is the example I use to introduce the topic. I extend the results to generalized Burgers' and KdV equations. I consider Matrix Hopf-Cole transformations and use this when I introduce Darboux transformations for linear diffusion and wave equations.

In Chap. 4, I introduce point and contact transformations. These are a generalization of the usual change of variables one would find in an introductory course in PDEs. I introduce three special transformations: the Hodograph transformation, the Legendre transformation, and the Ampere transformation. I also introduce conditions, the "contact conditions," which guarantee that a transformation will be a contact transformation. With several examples considered, the chapter ends with the famous Plateau problem from minimal surfaces.

In Chap. 5, we consider first integrals. Simply put, these are PDEs of a lower order (if they exist) and yield exact solutions of a given equation. I first consider second order quasilinear PDEs in two independent variables and then general Monge-Ampere (MA) equations in two independent variables. Often times, PDEs admit very general classes of first integrals. When this happens, it is sometimes possible to transform the given PDE to one that is very simple. I then consider three classes of MA equations: a class of hyperbolic, parabolic, and elliptic MA equations.

The final chapter, Chap. 6, is functional separability. This is a generalization of the usual separation of variables that one usually finds in an introductory course when solving PDEs like the heat equation, the wave equation, or Laplace's equation.

It is fair to say that this book does not cover all techniques used to solve nonlinear PDEs but rather represents techniques that I personally have interest in and have success-fully used. For example, techniques like the inverse scattering transform and Bäcklund transformations are extremely important and should be added to one's arsenal. It is also fair to say this is not the only book on techniques for solving PDEs. For example, one should also consider A. R. Forsyth's "A Treatise on Differential Equations," W. F. Ames' "Nonlinear Partial Differential Equations in Engineering, Vol. I and II," and L. Debnath's "Nonlinear Partial Differential Equations for Scientists and Engineers."

Conway, USA   Daniel Arrigo
April 2019

Acknowledgements

I first would like to thank my wife Peggy. She once again became a book widow. My love and thanks. Second, I would like to thank all of my students who, over the past 12+ years, volunteered to read the book and gave much needed input on both the presentation of material and on the complexity of the examples given. Finally, I would like to thank Susanne Filler of Springer Nature Publishers. Once again, she made the process a simple and straightforward one.

Contents

Nonlinear PDEs are Everywhere

Outside of Quantum mechanics, the world around us is modeled by nonlinear partial differential equations (PDEs). Here is just a short list of places that one may find nonlinear PDEs.

1. The nonlinear diffusion equation

$$u_t = (D(u)u_x)_x \tag{1.1}$$

is a nonlinear PDE that models heat transfer in a medium where the thermal conductivity may depend on the temperature. The equation also arises in numerous other fields such as soil physics, population genetics, fluid dynamics, neurology, combustion theory, and reaction chemistry, to name just a few (see [1] and the references within).

2. The nonlinear wave equation

$$u_{tt} = \left(c(u)^2 u_x\right)_x \tag{1.2}$$

essentially models wave propagation and appears in applications involving one-dimensional gases, shallow water waves, longitudinal threadlines, finite nonlinear strings, elastic-plastic materials, and transmission lines, to name a few (see [2] and the references within).

3. Burgers Equation

$$u_t + uu_x = \nu u_{xx} \tag{1.3}$$

is a partial differential equation that incorporates both nonlinearity and diffusion. It was first introduced as a simplified model for turbulence [3] and appears in various areas of applied mathematics, such as soil-water flow [4], nonlinear acoustics [5], and traffic flow [6].

4. Fisher's equation

$$u_t = u_{xx} + u(1 - u) \tag{1.4}$$

© The Author(s), under exclusive license to Springer Nature Switzerland AG 2022
D. Arrigo, *Analytical Methods for Solving Nonlinear Partial Differential Equations*,
Synthesis Lectures on Mathematics & Statistics.
https://doi.org/10.1007/978-3-031-17069-0_1

is a model proposed for the wave of advance of advantageous genes [7] and has applications in early farming [8], chemical wave propagation [9], nuclear reactors [10], chemical kinetics [11], and in theory of combustion [12].

5. The Fitzhugh-Nagumo equation

$$u_t = u_{xx} + u(1-u)(u+\lambda) \tag{1.5}$$

models the transmission of nerve impulses [13, 14] and arises in population genetics models [15].

6. The Korteweg deVries equation (KdV)

$$u_t + 6uu_x + u_{xxx} = 0 \tag{1.6}$$

describes the evolution of long water waves down a canal of rectangular cross section. It has also been shown to model longitudinal waves propagating in a one-dimensional lattice, ion-acoustic waves in a cold plasma, waves in elastic rods, and used to describe the axial component of velocity in a rotating fluid flow down a tube [16] .

7. The Boussinesq equation

$$u_{tt} + u_{xx} + (2uu_x)_x + \frac{1}{3}u_{xxxx} = 0 \tag{1.7}$$

was introduced by Boussinesq in 1871 [17, 18] to model shallow water waves on long channels. It also arises in other applications such as in one-dimensional nonlinear lattice waves [19, 20], vibrations in a nonlinear string [21], and ion sound waves in a plasma [22, 23].

8. The Eikonial equation

$$|\nabla u| = F(x), \quad x \in \mathbb{R}^n \tag{1.8}$$

appears in ray optics [24].

9. The Gross-Pitaevskii equation

$$i\psi_t = -\nabla^2\psi + (V(x) + |\psi|^2)\psi \tag{1.9}$$

is a model for the single-particle wavefunction in a Bose-Einstein condensate [25, 26].

10. Plateau's equation

$$(1 + u_y^2)u_{xx} - 2u_xu_yu_{xy} + (1 + u_x^2)u_{yy} = 0 \tag{1.10}$$

arises in the study of minimal surfaces [27].

11. The Sine-Gordon equation

$$u_{xy} = \sin u \tag{1.11}$$

arises in the study of surfaces of constant negative curvature [28], and in the study of crystal dislocations [29].

12. The equilibrium equations

$$\frac{\partial \sigma_{xx}}{\partial x} + \frac{\partial \sigma_{xy}}{\partial y} + F_x = 0$$

$$\frac{\partial \sigma_{xy}}{\partial x} + \frac{\partial \sigma_{yy}}{\partial y} + F_y = 0 \tag{1.12}$$

arise in elasticity. Here, σ_{xx}, σ_{xy} and σ_{yy} are normal and shear stresses, and F_x and F_y are body forces [30]. These have been used by Cox, Hill, and Thamwattana [31] (see also [32]) to model highly frictional granular materials.

13. The Navier-Stokes equations

$$\nabla \cdot \mathbf{u} = 0$$

$$\mathbf{u}_t + \mathbf{u} \cdot \nabla \mathbf{u} = -\frac{\nabla P}{\rho} + \nu \nabla^2 \mathbf{u} \tag{1.13}$$

describe the velocity field and pressure of incompressible fluids. Here ν is the kinematic viscosity, $\mathbf{u}$ is the velocity of the fluid parcel, P is the pressure, and ρ is the fluid density [33].

1.1 Exercises

1. Show solutions exist for the nonlinear diffusion equation

$$u_t = \left(u^m u_x\right)_x, \quad m \in \mathbb{R} \tag{1.14}$$

of the form $u = kt^p x^q$ for suitable constants k, p and q. Use these to obtain solutions to

$$u_t = (uu_x)_x \quad \text{and} \quad u_t = \left(\frac{u_x}{u}\right)_x. \tag{1.15}$$

2. Show that Fisher's equation

$$u_t = u_{xx} + u(1 - u) \tag{1.16}$$

admit solutions of the form $u = f(x - ct)$ where f satisfies the ordinary differential equation (ODE)

$$f'' + cf' + f - f^2 = 0. \tag{1.17}$$

Further show exact solutions can be obtained in the form

$$f = \frac{1}{\left(a + be^{kz}\right)^2} \tag{1.18}$$

for suitable constants a, b, c, and k [34].

3. Show that the Fitzhugh−Nagumo equation

$$u_t = u_{xx} + u(1 - u)(u + \lambda) \tag{1.19}$$

admits solutions of the form $u = f(x - ct)$ where f satisfies the ODE

$$f'' + cf' + f(1 - f)(f + \lambda) = 0. \tag{1.20}$$

Further show exact solutions can be obtained in the form

$$f = \frac{1}{a + be^{kz}} \tag{1.21}$$

for suitable constants a, b, c, λ and k.

4. Show that solutions exist of the form

$$u = \frac{ax}{x^2 + bt}$$

(a and b constant) that satisfies Burgers' equation (1.3).

5. Consider the PDE

$$u_t = u_{xx} + 2 \operatorname{sech}^2 x\, u. \tag{1.22}$$

Even though linear, exact solutions to equations of the form

$$u_t = u_{xx} + f(x)u \tag{1.23}$$

can be difficult to find. If $v = e^{k^2 t} \sinh kt$ or $v = e^{k^2 t} \cosh kt$ (k is an arbitrary constant), show

$$u = v_x - \tanh x \cdot v$$

satisfies (1.22).

6. Show

$$u = \ln \left| \frac{2f'(x)g'(y)}{(f(x) + g(y))^2} \right|$$

where $f(x)$ and $g(y)$ are arbitrary functions satisfies Liouville's equation

$$u_{xy} = e^u.$$

7. Show

$$u = 4\tan^{-1}\left(e^{ax+a^{-1}y}\right)$$

where a is an arbitrary nonzero constant satisfies the Sine-Gordon equation

$$u_{xy} = \sin u.$$

8. Show that if

$$u = f(x + ct)$$

satisfies the KdV equation (1.6) then f satisfies

$$cf' + 6ff' + f''' = 0 \tag{1.24}$$

where prime denotes differentiation with respect to the argument of f. Show there is one value of c such that $f(r) = 2\,\text{sech}^2 r$ is a solution of (1.24).

9. The PDE

$$v_t - 6v^2 v_x + v_{xxx} = 0$$

is known as the modified Korteweg de Vries (mKdV) equation. Show that if v is a solution of the mKdV, then

$$u = v_x - v^2$$

is a solution of the KdV (1.6).

10. The Boussinesq equation

$$u_{tt} + u_{xx} - \left(u^2\right)_{xx} - \frac{1}{3}u_{xxxx} = 0 \tag{1.25}$$

under the substitution $u = 2\,(\ln F(t, x))_{xx}$ (and integrating twice) becomes ([35])

$$FF_{tt} - F_t^2 + FF_{xx} - F_x^2 - \frac{1}{3}\left(FF_{xxxx} - 4F_x F_{xxx} + 3F_{xx}^2\right) = 0. \tag{1.26}$$

Further show that (1.26) admits solutions of the form $F = ax^2 + bt^2 + c$ for suitable a, b and c (see [36]).

References

1. B.H. Gilding, R. Kersner, The characterization of reaction-convection-diffusion processes by travelling waves. J. Differ. Equ. **124**, 27–79 (1996)
2. W.F. Ames, R.J. Lohner, E. Adams, Group Properties of $u_{tt} = [f(u)u_x]_x$ Int. J. Nonlin. Mech. **16**(5/6), 439–447 (1981)
3. J.M. Burgers, Mathematical examples illustrating relations occuring in the theory of turbulent fluid motion. Akademie van Wetenschappen, Amsterdam, Eerste Sectie, Deel XVII, No. **2**, 1–53 (1939)
4. P. Broadbridge, Burgers' equation and layered media: exact solutions and applications to soil-water flow. Math. Comput. Model. **16**(11), 163–169 (1992)
5. K. Naugolnykh, L. Ostrovsky, *Nonlinear Wave Processes in Acoustics* (Cambridge University Press, 1998)
6. R. Haberman, *Mathematical Models; Mechanical Vibrations, Population Dynamics, and Traffic Flow: An Introduction to Applied Mathematics*, Imprint Englewood Cliffs (Prentice-Hall, N.J., 1977)
7. R.A. Fisher, The wave of advance of advantageous genes. Ann. Eugen. **7**, 353–369 (1937)
8. A.J. Ammermann, L.L. Cavalli-Sforva, Measuring the rate of spread of early farming. Man **6**, 674–688 (1971)
9. R. Arnold, K. Showalter, J.J. Tyson, Propagation of chemical reactions in space. J. Chem. Educ. **64**, 740–742 (1987)
10. J. Canosa, Diffusion in nonlinear multiplicative media. J. Math. Phys. **10**, 1862–1868 (1969)
11. P.C. Fife, *Mathematical Aspects of Reacting and Diffusing Systems*, Lectures in Biomathematics, vol. 28 (Springer, Berlin, 1979)
12. A. Kolmogorov, I. Petrovsky, N. Piscunov, A study of the equation of diffusion with increase in the quantity of matter and its application to a biological problem. Bull. Univ. Moscow, Ser. Int. Sec A. **1**, 1–25 (1937)
13. R. Fitzhugh, Impulses and physiological states in theoretical models of nerve membrane. Biophys. J. **1**, 445–466 (1961)
14. J.S. Nagumo, S. Arimoto, S. Yoshizawa, An active pulse transmission line simulating nerve axon. Proc. IRE **50**, 2061–2070 (1962)
15. D.G. Aronson, H.F. Weinberger, in *Partial Differential Equations and Related Topics*, ed. by J.A. Goldstein, Lecture Notes in Mathematics, vol. 446. (Springer, Berlin, 1975)
16. R. Miura, The Korteweg–deVries equation: a survey of results. SIAM Rev. **18**(3), 412–459 (1978)
17. J. Boussinesq, Theorie de l'intumescence liquide, appelee onde solitaire ou de translation se propagente dans un canal rectangulaire. Comptes Rendus **72**, 755–759 (1871)
18. J. Boussinesq, Theorie des ondes et des remous qui se propagent le long d'un canal rectangulaire horizontal, en communi- quant au liquide contenu dans ce canal des vitesses sensiblemant parielles de la surface au fond. J. Pure Appl. **17**, 55–108 (1872)
19. M. Toda, Studies of a nonlinear lattice. Phys. Rep. **8**, 1–125 (1975)
20. N.J. Zabusky, A synergetic approach to problems of nonlinear dispersive wave propagation and interaction, in *Nonlinear Partial Differential Equations* ed. by W.F. Ames (Academic, New York, 1967), pp. 233–258
21. V.E. Zakharov, On stocastization of one-dimensional chains of nonlinear oscillations. Sov. Phys. JETP **38**, 108–110 (1974)
22. E. Infeld, G. Rowlands, *Nonlinear Waves, Solitons and Chaos* (C.U.P., Cambridge, 1990)

23. A.C. Scott, The application of Backlund transforms to physical problems, in *Backlund Transformations* ed. by R.M. Miura, Lecture Notes in Mathematics, vol. 515 (Springer, Berlin, 1975), pp. 80–105
24. D.D. Holm, *Geometric Mechanics Part 1: Dynamics and Symmetry* (Imperial College Press, 2011)
25. E.P. Gross, Structure of a quantized vortex in boson systems. Il Nuovo Cimento. **20**(3), 454–457 (1961)
26. L.P. Pitaevskii, Vortex lines in an imperfect Bose gas. Sov. Phys. JETP. **13**(2), 451–454 (1961)
27. J.C.C. Nitsche, On new results in the theory of minimal surfaces. Bull. Am. Math. Soc. **71**, 195–270 (1965)
28. L.P. Eisenhart, *A Treatise on the Differential Geometry of Curves and Surfaces* (Dover Publ, New York, 1960)
29. F.C. Frank, J.H. van der Merwe, One-dimensional dislocations. I. Static theory. Proc. R. Soc. Lond. A **198**, 205–216 (1949)
30. S. Timoshenko, J.N. Goodier, *Theory of Elasticity* (McGraw-Hill, 1951)
31. G.M. Cox, J.M. Hill, N. Thamwattana, A formal exact mathematical solution for a sloping rat-hole in a highly frictional granular solid. Acta Mech. **170**, 127–147 (2004)
32. D.J. Arrigo, L. Le, J.W Torrence, Exact solutions for a class of ratholes in highly frictional granular solids. Dyn. Cont., Dis. Imp. Sys. B: Appl. Algor. **19**, 497–509 (2012)
33. G.K. Batchelor, *An introduction to Fluid Mechanics* (Cambridge University Press, 2000)
34. M.J. Ablowitz, A. Zeppetella, Explicit solutions of Fisher's equations for a special wave speed. Bull. Math. Bio. **41**, 835–840 (1979)
35. R. Hirota, Exact N-soliton solutions of the wave equation of long waves in shallow-water and in nonlinear lattices. J. Math. Phys. **14**, 810–814 (1973)
36. M.J. Ablowitz, J. Satsuma, Solitons and rational solutions of nonlinear evolution equations. J. Math. Phys. **19**, 2180–2186 (1978)

Compatibility

We start our discussion by first solving the nonlinear PDE

$$xu_x - u_y^2 = 2u, \tag{2.1}$$

subject to the boundary condition

$$u(x, x) = 0. \tag{2.2}$$

As with most introductory courses in partial differential equations (see, for example, [1]) we use the method of characteristics. Here, we define F as

$$F = xp - q^2 - 2u. \tag{2.3}$$

The characteristic equations become

$$x_s = F_p = x, \tag{2.4a}$$
$$y_s = F_q = -2q, \tag{2.4b}$$
$$u_s = pF_p + qF_q = xp - 2q^2, \tag{2.4c}$$
$$p_s = -F_x - pF_u = p, \tag{2.4d}$$
$$q_s = -F_y - qF_u = 2q. \tag{2.4e}$$

In order to solve the PDE (2.1) we will need to solve the system (2.4). As (2.1) has a boundary condition (BC), we will create BCs for the system (2.4). In the (x, y) plane, the line $y = x$ is the boundary where u is defined. To this, we associate a boundary in the (r, s) plane. Given the flexibility, we can choose $s = 0$ and connect the two boundaries via $x = r$ (Fig. 2.1). Therefore, we have

$$x = r, \quad y = r, \quad u = 0 \text{ when } s = 0. \tag{2.5}$$

© The Author(s), under exclusive license to Springer Nature Switzerland AG 2022
D. Arrigo, *Analytical Methods for Solving Nonlinear Partial Differential Equations,*
Synthesis Lectures on Mathematics & Statistics.
https://doi.org/10.1007/978-3-031-17069-0_2

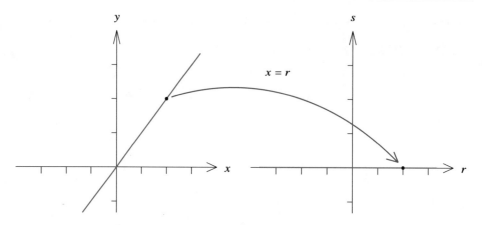

Fig. 2.1 Change in the boundary from the (x, y) plane to the (r, s) plane

To determine p and q on $s = 0$, it is necessary to consider the initial condition $u(x, x) = 0$. Differentiating with respect to x gives

$$u_x(x, x) + u_y(x, x) = 0. \tag{2.6}$$

From the original PDE (2.1)

$$x u_x(x, x) - u_y^2(x, x) = 0. \tag{2.7}$$

If we denote $p_0 = u_x(x, x)$ and $q_0 = u_y(x, y)$, then (2.6) and (2.7) become

$$p_0 + q_0 = 0, \quad r p_0 - q_0^2 = 0. \tag{2.8}$$

These are easily solved giving

$$(i) \quad p_0 = 0, \quad q_0 = 0, \tag{2.9a}$$
$$(ii) \quad p_0 = r, \quad q_0 = -r. \tag{2.9b}$$

As case (i) gives rise to the trivial solution $u = 0$, we will only consider case (ii).
    Solving (2.4a) gives

$$x = a(r)e^s, \tag{2.10}$$

where a is an arbitrary function. From the boundary condition (2.5), we find that $a(r) = r$ gives

$$x = re^s. \tag{2.11}$$

Solving (2.4d) gives

$$p = b(r)e^s, \tag{2.12}$$

where b is an arbitrary function. From the boundary condition (2.9b), we find that $b(r) = r$ gives

$$p = re^s. \tag{2.13}$$

Solving (2.4e) gives

$$q = c(r)e^{2s}, \tag{2.14}$$

where c is an arbitrary function. From the boundary condition (2.9b), we find that $c(r) = -r$ gives

$$q = -re^{2s}. \tag{2.15}$$

To solve (2.4b) we need to use q. Thus,

$$y_s = -2q = 2re^{2s} \tag{2.16}$$

easily integrates, giving $y = re^{2s} + d(r)$. Using the BC (2.5) shows that $d = 0$ gives

$$y = re^{2s}. \tag{2.17}$$

Finally, we see from (2.4b)

$$u_s = xp - 2q^2 = r^2 e^{2s} - 2r^2 e^{4s}, \tag{2.18}$$

which again integrates easily giving $u = \frac{1}{2}(r^2 e^{2s} - r^2 e^{4s}) + e(r)$, where e is an arbitrary function. From the boundary condition (2.5), we find that $e(r) = 0$ gives

$$u = \frac{r^2 e^{2s} - r^2 e^{4s}}{2}. \tag{2.19}$$

Eliminating r and s from (2.11), (2.17), and (2.19) gives

$$u = \frac{x^2 - y^2}{2}. \tag{2.20}$$

The reader can verify the the solution (2.20) does indeed satisfy the PDE (2.1) and the BC (2.2).

2.1   Charpit's Method

Obtaining exact solutions to nonlinear PDEs such as (2.1) can be a difficult task as we are required to solve equations such as (2.4)! The difficulty is not so much in solving these characteristic equations but in eliminating the 5 arbitrary functions that appear upon integration. So we ask, is it possible to obtain exact solutions another way?

Consider the PDE

$$u_y = -y. \tag{2.21}$$

This integrates to give

$$u = -\frac{y^2}{2} + f(x), \tag{2.22}$$

where f is an arbitrary function of integration and substitution into the original PDE (2.1) gives

$$xf' = 2f. \tag{2.23}$$

This ODE is solved giving

$$f = cx^2, \tag{2.24}$$

which from (2.22) leads to

$$u = cx^2 - \frac{y^2}{2}. \tag{2.25}$$

The initial condition (2.2) gives $c = 1/2$, and leads to the solution given in (2.20).

Consider the PDE

$$u_x = x. \tag{2.26}$$

This integrates to give

$$u = \frac{x^2}{2} + g(y), \tag{2.27}$$

where g is an arbitrary function of integration and substitution into the original PDE (2.1) gives

$$-g'^2 = 2g. \tag{2.28}$$

This ODE is solved giving (we will omit the trivial solution $g = 0$)

$$g = -\frac{y^2}{2} + cy - \frac{c^2}{2}, \tag{2.29}$$

which from (2.27) and the initial condition (2.2) again leads to the solution (2.20).

As for the final example, consider the PDE

$$u_x + xu_y = x - xy. \tag{2.30}$$

This integrates to give

$$u = y - \frac{1}{2}y^2 + h\left(y - \frac{1}{2}x^2\right), \tag{2.31}$$

where h is an arbitrary function of its argument. Substitution into the original PDE (2.1) gives

$$2(\lambda - 1)h' - h'^2 - 1 = 2h, \tag{2.32}$$

where $h = h(\lambda)$ and $\lambda = y - \frac{1}{2}x^2$. This ODE actually has two solutions

$$h = c\lambda - \frac{1}{2}(c+1)^2, \quad h = \frac{1}{2}\lambda^2 - \lambda, \tag{2.33}$$

and leads to the exact solutions

$$u = y - \frac{1}{2}^2 + c\left(y - \frac{1}{2}x^2\right) - \frac{1}{2}(c+1)^2,$$
$$u = \frac{1}{2}(x^2 - y^2) + \frac{1}{2}\left(y - \frac{1}{2}x^2\right)^2. \tag{2.34}$$

However, only the first solution will give rise to (2.20).

A question we ask: Is it really necessary to solve the reduced ODEs given in (2.23), (2.28), and (2.32)? Consider the solution forms that lead to these ODEs

$$u = -\frac{y^2}{2} + f(x),$$
$$u = \frac{x^2}{2} + g(y), \tag{2.35}$$
$$u = y - \frac{1}{2}y^2 + h\left(y - \frac{1}{2}x^2\right).$$

If we impose the initial condition $u(x, x) = 0$ in each solution form (2.35), we obtain the solution $u = \frac{1}{2}(x^2 - y^2)$, but we certainly would have missed obtaining the second exact solution in (2.34). So are there others? For example, both

$$u_y = x^2,$$
$$xu_x + 2yu_y = 2u - y^2, \tag{2.36}$$

will lead to exact solutions of the given PDE (2.1) however, the boundary conditions might not be satisfied. In fact any PDE of the form

$$F\left(\frac{u_x}{x}, \frac{u_x^2}{u_y}, u_y + y, xu_x - u_y^2 - 2u\right) = 0 \tag{2.37}$$

will give rise to an exact solution to (2.1). A number of questions arise.

1. Where did these associated PDEs come from?

2. How do we know that they will lead to a solution that also satisfies the BC?

Before trying to answer such questions, it is important to know that an actual solution exists. Namely, does a solution exist that satisfies both the original PDE and second appended PDE? So, in the first example, does a solution exist that satisfies both

$$xu_x - u_y^2 = 2u \quad \text{and} \quad u_y = -y? \tag{2.38}$$

If we isolate u_x and u_y in (2.38) and ask: does a solution exist to

$$u_x = \frac{2u}{x} + \frac{y^2}{x}, \quad u_y = -y? \tag{2.39}$$

If so, then they certainly would be compatible, so $\dfrac{\partial u_x}{\partial y} = \dfrac{\partial u_y}{\partial x}$. Calculating these gives

$$\frac{2u_y}{x} + \frac{2y}{x} \overset{?}{=} 0, \tag{2.40}$$

and since $u_y = -y$, then (2.40) is identically satisfied, so the two equations (2.38) are compatible. For the second example we ask: are the following compatible?

$$xu_x - u_y^2 = 2u \quad \text{and} \quad u_x = x. \tag{2.41}$$

We substitute $u_x = x$ into the first of (2.41) and then seek compatibility of

$$x^2 - u_y^2 = 2u \quad \text{and} \quad u_x = x. \tag{2.42}$$

To show compatibility, we differentiate the first of (2.42) with respect to x giving

$$2x - 2u_y u_{xy} = 2u_x, \tag{2.43}$$

and since $u_x = x$ then (2.43) is identically satisfied then (2.41) are compatible. For the final example, we ask: are these compatible?

$$xu_x - u_y^2 = 2u \quad \text{and} \quad u_x + xu_y = x - xy. \tag{2.44}$$

Definitely a harder problem to explicitly find u_x and u_y but is that really necessary? If they were compatible, they would have the same second derivatives. If we calculate the x and y derivatives of each we obtain

$$xu_{xx} - 2u_y u_{xy} = u_x, \tag{2.45a}$$
$$xu_{xy} - 2u_y u_{yy} = 2u_y, \tag{2.45b}$$
$$u_{xx} + xu_{xy} + u_y = 1 - y, \tag{2.45c}$$
$$u_{xy} + xu_{yy} = -x. \tag{2.45d}$$

Solving (2.45c) and (2.45d) for u_{xx} and u_{xy} gives

$$u_{xx} = x^2 u_{yy} - u_y + x^2 - y + 1, \quad u_{xy} = -x u_{yy} - x, \qquad (2.46)$$

and the two remaining equation in (2.45) become

$$x(2u_y + x^2)u_{yy} - u_x + x u_y + x^3 - xy + x = 0, \qquad (2.47a)$$

$$(2u_y + x^2)(u_{yy} + 1) = 0. \qquad (2.47b)$$

If (2.44) were to be compatible, then (2.47) would be identically satisfied. So we see two cases emerge:

$$\text{(i) } 2u_y + x^2 = 0,$$
$$\text{(ii) } 2u_y + x^2 \neq 0.$$

In the first case where $u_y = \frac{1}{2}x^2$, (2.47a) reduces to

$$2u_x - \frac{1}{3}x^3 - x + xy = 0,$$

which is compatible with $u_y = \frac{1}{2}x^2$ and so (2.44) are compatible. In the second case where $u_{yy} = -1$, (2.47a) reduces to

$$u_x + x u_y + x(y - 1) = 0$$

which is identically satisfied by virtue of (2.44) and again, (2.44) are compatible.

So now we know how to determine when two first order PDEs are compatible. Our next step is to determine how they come about.

Consider the compatibility of the following first order PDEs

$$F(x, y, u, p, q) = 0,$$
$$G(x, y, u, p, q) = 0, \qquad (2.48)$$

where $p = u_x$ and $q = u_y$. Calculating the x and y derivatives of (2.48) gives

$$\begin{aligned}
F_x + p F_u + u_{xx} F_p + u_{xy} F_q &= 0, \\
F_y + q F_u + u_{xy} F_p + u_{yy} F_q &= 0, \\
G_x + p G_u + u_{xx} G_p + u_{xy} G_q &= 0, \\
G_y + q G_u + u_{xy} G_p + u_{yy} G_q &= 0.
\end{aligned} \qquad (2.49)$$

Solving the first three (2.49) for u_{xx}, u_{xy} and u_{yy} gives

$$u_{xx} = \frac{-F_x\, G_q - p\, F_u\, G_q + F_q\, G_x + p\, F_q\, G_u}{F_p\, G_q - F_q\, G_p},$$

$$u_{xy} = \frac{-F_p\, G_x - p\, F_p\, G_u + F_x\, G_p + p\, F_u\, G_p}{F_p\, G_q - F_q\, G_p},$$

$$u_{yy} = \frac{\begin{array}{l} F_p^2\, G_x + p\, F_p^2\, G_u - F_y\, F_p\, G_q - q\, F_u\, F_p\, G_q \\ + q\, F_u\, F_q\, G_p - F_x\, F_p\, G_p - p\, F_u\, F_p\, G_p + F_y\, F_q\, G_p \end{array}}{(F_p\, G_q - F_q\, G_p)F_q}.$$

Substitution into the last of (2.49) gives

$$F_p\, G_x + F_q\, G_y + (p\, F_p + q\, F_q)G_u - (F_x + p\, F_u)G_p - (F_y + q\, F_u)G_q = 0,$$

conveniently written as

$$\begin{vmatrix} D_x F & F_p \\ D_x G & G_p \end{vmatrix} + \begin{vmatrix} D_y F & F_q \\ D_y G & G_q \end{vmatrix} = 0, \tag{2.51}$$

where $D_x\, F = F_x + p\, F_u$, $D_y\, F = F_y + q\, F_u$ and $|\cdot|$ is the usual determinant. These are known as the *Charpit equations*. We also assumed that $F_p G_q - F_q G_p \neq 0$ and $F_q \neq 0$. These cases would need to be considered separately.

Example 2.1 Consider

$$x u_x - u_y^2 = 2u. \tag{2.52}$$

This is the example we already considered; now we will determine all classes of equations that are compatible with this one. Denoting

$$G = x u_x - u_y^2 - 2u$$
$$= xp - q^2 - 2u,$$

where $p = u_x$ and $q = u_y$, then

$$G_x = p, \quad G_y = 0, \quad G_u = -2, \quad G_p = x, \quad G_q = -2q,$$

and the Charpit equations are

$$\begin{vmatrix} D_x F & F_p \\ -p & x \end{vmatrix} + \begin{vmatrix} D_y F & F_q \\ -2q & -2q \end{vmatrix} = 0,$$

or, after expansion,

$$x F_x - 2q\, F_y + (xp - 2q^2)\, F_u + p F_p + 2q F_q = 0.$$

Solving this linear PDE by the method of characteristics gives the solution as

$$F\left(\frac{p}{x}, \frac{p^2}{q}, q + y, xp - q^2 - 2u\right) = 0 \tag{2.53}$$

which is exactly the one given in (2.37)!

Example 2.2 Consider

$$u_x u_y = 1. \tag{2.54}$$

Denoting

$$G = u_x u_y - 1 = pq - 1,$$

where $p = u_x$ and $q = u_y$, then

$$G_x = 0, \quad G_y = 0, \quad G_u = 0, \quad G_p = q, \quad G_q = p,$$

and the Charpit equations are

$$\begin{vmatrix} D_x F & F_p \\ 0 & q \end{vmatrix} + \begin{vmatrix} D_y F & F_q \\ 0 & -p \end{vmatrix} = 0,$$

or, after expansion,

$$q F_x + p F_y + 2pq F_u = 0.$$

Note that the third term can be replaced by $2F_u$, due to the original equation. Solving this linear PDE by the method of characteristics gives the solution as

$$F = F(uq - 2x, up - 2y, p, q), \tag{2.55}$$

or

$$F = F(uu_y - 2x, uu_x - 2y, u_x, u_y). \tag{2.56}$$

For example, if we choose

$$uu_y - 2x = 0, \tag{2.57}$$

then on integrating we obtain

$$u^2 = 4xy + f(x), \tag{2.58}$$

where f is an arbitrary function of integration and substituting into the original PDE (2.54) gives

$$xf' = f. \tag{2.59}$$

This ODE is easily integrated giving

$$f = cx, \tag{2.60}$$

where c is a constant of integration and leads to the exact solution

$$u^2 = 4xy + cx. \tag{2.61}$$

If we choose

$$uu_x + u_y - 2y = 0, \tag{2.62}$$

then on integrating we obtain

$$u - y^2 + f\left(yu - x - \frac{2}{3}y^3\right) = 0, \tag{2.63}$$

where f is an arbitrary function of its argument and substituting into the original PDE (2.54) gives

$$ff'^2 - 1 = 0. \tag{2.64}$$

Again, this ODE is easily integrated giving

$$f = \left(c \pm \frac{3}{2}x\right)^{3/2} \tag{2.65}$$

and leads to the exact solution

$$u - y^2 + \left(c \pm \frac{3}{2}(yu - x - \frac{2}{3}y^3)\right)^{2/3} = 0. \tag{2.66}$$

Example 2.3 Consider

$$u_x^2 + u_y^2 = u^2. \tag{2.67}$$

Denoting $p = u_x$ and $q = u_y$, then

$$G = u_x^2 + u_y^2 - u^2 = p^2 + q^2 - u^2.$$

Thus,

$$G_x = 0, \quad G_y = 0, \quad G_u = -2u, \quad G_p = 2p, \quad G_q = 2q,$$

and the Charpit equation's are

$$\begin{vmatrix} D_x F & F_p \\ -2pu & 2p \end{vmatrix} + \begin{vmatrix} D_y F & F_q \\ -2qu & 2q \end{vmatrix} = 0,$$

or, after expansion,

$$pF_x + qF_y + \left(p^2 + q^2\right)F_u + puF_p + quF_q = 0. \tag{2.68}$$

Note that the third term can be replaced by $u^2 F_u$, due to the original equation. Solving (2.68), a linear PDE, by the method of characteristics gives the solution as

$$F = F\left(x - \frac{p}{u}\ln u, \, y - \frac{q}{u}\ln u, \, \frac{p}{u}, \frac{q}{u}\right).$$

One particular example is

$$x - \frac{p}{u} \ln u + y - \frac{q}{u} \ln u = 0,$$

or

$$u_x + u_y = (x + y) \frac{u}{\ln u}.$$

If we let $u = e^{\sqrt{v}}$, this becomes

$$v_x + v_y = 2(x + y),$$

which, by the method of characteristics, has the solution

$$v = 2xy + f(x - y).$$

This, in turn, gives the solution for u as

$$u = e^{\sqrt{2xy + f(x-y)}}. \tag{2.69}$$

Substitution into the original Eq. (2.67) gives the following ODE

$$f'^2 - 2\lambda f' - 2f + 2\lambda^2 = 0,$$

where $f = f(\lambda)$ and $\lambda = x - y$. If we let $f = g + \frac{1}{2}\lambda^2$ then we obtain

$$g'^2 - 2g = 0 \tag{2.70}$$

whose solutions are given by

$$g = \frac{(r + c)^2}{2}, \quad g = 0, \tag{2.71}$$

where c is an arbitrary constant of integration. This, in turn, gives

$$f = \lambda^2 + c\lambda + \frac{1}{2}c^2, \quad f = \frac{1}{2}\lambda^2 \tag{2.72}$$

and substitution into (2.69) gives

$$u = e^{\sqrt{x^2 + y^2 + c(x-y) + \frac{1}{2}c^2}}, \quad u = e^{\sqrt{2xy + (x-y)^2/2}},$$

exact solutions to the original PDE (2.67).

It is interesting to note that when we substitute the solution of the compatible equation into the original it reduces to an ODE. A natural question is: does this always happen? This was proven to be true in two independent variables by the author [2].

So now we ask, of the infinite possibilities of compatible equations, can we choose the right one(s) to not only solve a given PDE but also satisfy the given boundary condition (BC)? The following example illustrates.

Example 2.4 Solve

$$u_x u_y - x u_x - y u_y = 0, \tag{2.73}$$

subject to the boundary conditions

$$
\begin{aligned}
&(i) \ \ u(x, 1) = 0, \\
&(ii) \ \ u(x, 0) = \frac{1}{2}x^2, \\
&(iii) \ \ u(x, x) = 2x^2.
\end{aligned}
\tag{2.74}
$$

Denoting

$$G = pq - xp - yq,$$

where $p = u_x$ and $q = u_y$, then

$$G_x = -p, \ \ G_y = -q, \ \ G_u = 0, \ \ G_p = q - x, \ \ G_q = p - y,$$

and the Charpit equations are

$$
\begin{vmatrix} D_x F & F_p \\ -p & q - x \end{vmatrix} + \begin{vmatrix} D_y F & F_q \\ -q & p - y \end{vmatrix} = 0,
$$

or, after expansion,

$$(q - x)F_x + (p - y)F_y + (2pq - xp - yq)F_u + pF_p + qF_q = 0. \tag{2.75}$$

Note that the third term can be replaced by $pq F_u$, due to the original PDE. Solving (2.75) by the method of characteristics gives the solution as

$$F = F(q^2 - 2xq, \ p^2 - 2yp, \ p/q, \ pq - 2u), \tag{2.76}$$

or

$$F = F(u_y^2 - 2xu_y, \ u_x^2 - 2yu_x, \ u_x/u_y, \ u_x u_y - 2u). \tag{2.77}$$

So how do we incorporate the BCs? Here, we will look at each BC separately.

Boundary Condition (i) In this case $u(x, 1) = 0$, and differentiating with respect to x gives $u_x(x, 1) = 0$ on the boundary. From the original PDE (2.73) we then have $u_y(x, 1) = 0$. Substituting these into (2.77) gives

$$F = F(0, 0, ?, 0). \tag{2.78}$$

Note that we have included a ? in the third argument of F since we have $\frac{0}{0}$. So how do we now use (2.78)? If we choose any of the arguments in (2.78) that are zero, then the boundary conditions are satisfied by that particular PDE. So, if we choose the first for example, we have

$$u_y^2 - 2xu_y = 0 \tag{2.79}$$

and we know that this is compatible with the original PDE (2.73) and satisfies the BC (2.74) and will give rise to the desired solution. It appears there are two cases (a) $u_y = 0$ and (b) $u_y - 2x = 0$, however the second case cannot be satisfied on the boundary as $u_y = 0$ gives $-2x = 0$ (inadmissible). Thus, from case (a) we have $u_y = 0$ and from the original PDE $u_x = 0$ giving $u = c$ and the BC $u(x, 1) = 0$ gives $c = 0$, so the solution is $u \equiv 0$.

Boundary Condition (ii) In this case $u(x, 0) = \frac{1}{2}x^2$, and differentiating with respect to x gives $u_x(x, 0) = x$ on the boundary. From the original PDE we then have $xu_y(x, 0) - x^2 = 0$ or $u_y(x, 0) = x$. Substituting these into (2.77) gives

$$F = F(-x^2, x^2, 1, 0). \tag{2.80}$$

So how do we use (2.80)? Again, if we choose any combination of the arguments that is zero, then the boundary conditions are satisfied by that particular PDE. Thus, if we choose the sum of the first two arguments i.e. $u_y^2 - 2xu_y + u_x^2 - 2yu_x = 0$ then the solution of this will satisfy the BC $u(x, y) = \frac{1}{2}x^2$. However, this PDE is nonlinear. We would like to solve a linear problem if we can so we will choose a different combination. Another choice would be choosing the third argument in (2.80) equals 1, i.e. $u_x/u_y = 1$. So we are to solve

$$u_x - u_y = 0. \tag{2.81}$$

This is easily solved giving $u = f(x + y)$ and substitution into the original PDE gives

$$f'^2 - \lambda f' = 0 \quad \text{or} \quad f' = 0, \quad f' - \lambda = 0, \tag{2.82}$$

where $\lambda = x + y$. Only the second gives rise to a correct solution and we find that $f = \frac{1}{2}\lambda^2$ and leads to the exact solution $u = \frac{1}{2}(x + y)^2$.

Boundary Condition (iii) In this case $u(x, x) = 2x^2$, and differentiating with respect to x gives $u_x(x, x) + u_y(x, y) = 4x$ on the boundary. From the original PDE we then have $u_x(x, x)u_y(x, x) - xu_x(x, x) - xu_y(x, x) = 0$. Solving for $u_x(x, x)$ and $u_y(x, x)$ gives $u_x(x, x) = u_y(x, x) = 2x$. Substituting these into (2.77) gives

$$F = F(0, 0, 1, 0). \tag{2.83}$$

So now we have lots of possibilities Again, if we choose any combination of the arguments that is zero, then the boundary conditions are satisfied by that particular PDE. For example,

$$(a)\ u_y^2 - 2xu_y = 0,$$
$$(b)\ u_x^2 - 2yu_x = 0, \tag{2.84}$$
$$(c)\ u_x/u_y - 1 = 0,$$

will all leads to solutions that satisfy the BC. PDE (a) leads to $u = 2xy$, PDE (b) leads to the same solution, and PDE (c) leads to $u = \frac{1}{2}(x + y)^2$. As another possibility, consider

$$F(a, b, c, d) = \frac{1}{2}\left(d - \frac{b}{c}\right).$$

This leads to

$$yu_y = u \tag{2.85}$$

which is easily solved, giving $u = yf(x)$. Substitution into the original PDE (2.73) gives

$$ff' - xf' - f = 0. \tag{2.86}$$

Now (2.86) is nonlinear and can be integrated. Sometimes the solution form and boundary condition give us a hint of what solution to look for. Since $u(x, x) = 2x^2$, using $u = yf(x)$ gives $xf(x) = 2x^2$ or $f(x) = 2x$, which in fact does satisfy the ODE (2.86)!

Up to this point, we have considered the compatibility of two first order PDEs. We now extend this technique to higher order PDEs.

2.2   Second Order PDEs

Consider the following pairs of PDEs

$$u_t + u_x = x, \quad u_t = u_{xx} \tag{2.87}$$

and

$$u_t + u_x = t, \quad u_t = u_{xx}. \tag{2.88}$$

We ask: does a common solution exist for each pair, (2.87) and (2.88)? We certainly could solve each of the first order PDEs in (2.87) and (2.88) and substitute their solutions into the second equation of (2.87) and (2.88), respectively and determine whether solutions exist, but instead we will ask if they are compatible. Compatible in the sense that all higher order derivatives would be the same.

Let us consider the pair (2.87). If we isolate the derivatives u_t and u_{xx}, then

$$u_t = x - u_x, \quad u_{xx} = x - u_x. \tag{2.89}$$

If these are to be compatible, then $(u_t)_{xx} = (u_{xx})_t$. Using (2.89), this compatibility constraint becomes

$$(x - u_x)_{xx} = (x - u_x)_t \quad \text{or} \quad u_{xxx} = u_{tx} \tag{2.90}$$

and, by virtue of the diffusion equation in (2.87), (2.90) is identically satisfied and thus, the pair (2.87) are compatible.

For the second pair, (2.88), isolating the derivatives u_t and u_{xx} gives

$$u_t = t - u_x, \quad u_{xx} = t - u_x. \tag{2.91}$$

If these are to be compatible, then $(u_t)_{xx} = (u_{xx})_t$. Therefore

$$(t - u_x)_{xx} = (t - u_x)_t \quad \text{or} \quad -u_{xxx} = 1 - u_{tx}, \tag{2.92}$$

and, by virtue of the diffusion equation in (2.88), gives $0 = 1$ which clearly is not true, so (2.88) are not compatible. As a matter of exercise, let us determine functions $f(t, x)$ such that the following are compatible

$$u_t + u_x = f(t, x), \quad u_t = u_{xx}. \tag{2.93}$$

As we did previously, we isolate the derivatives u_t and u_{xx}. This gives

$$u_t = f(t, x) - u_x, \quad u_{xx} = f(t, x) - u_x. \tag{2.94}$$

If these are to be compatible, then $(u_t)_{xx} = (u_{xx})_t$. This gives

$$(f(t, x) - u_x)_{xx} = (f(t, x) - u_x)_t \quad \text{or} \quad f_{xx} - u_{xxx} = f_t - u_{tx} \tag{2.95}$$

and, by virtue of the diffusion equation in (2.93), gives $f_{xx} = f_t$. Thus, if f is any solution of the diffusion equation, then (2.93) will be compatible. So how complicated can this get? The next example illustrates.

Determine functions f and g such that the following are compatible.

$$u_t + f(t, x)u_x = g(t, x)u, \quad u_t = u_{xx}. \tag{2.96}$$

We first isolate the derivatives u_t and u_{xx}. This gives

$$u_t = g(t, x)u - f(t, x)u_x, \quad u_{xx} = g(t, x)u - f(t, x)u_x. \tag{2.97}$$

Imposing $(u_t)_{xx} = (u_{xx})_t$ gives

$$g_{xx}u + 2g_x u_x + g u_{xx} - f_{xx}u_x - 2f_x u_{xx} - f u_{xxx} = g_t u + g u_t - f_t u_x - f u_{tx}. \tag{2.98}$$

Using (2.97) to eliminate the derivatives u_t, u_{tx}, u_{xx} and u_{xxx} gives

$$(g_{xx} - 2g f_x - g_t) u + (2g_x - f_{xx} + 2f f_x + f_t) u_x = 0. \tag{2.99}$$

For (2.96) to be compatible, then (2.99) this would have to be identically zero and, since u can vary, then the coefficients of u and u_x in (2.99) must be zero. This means that f and g must satisfy

$$g_t + 2gf_x - g_{xx} = 0, \qquad (2.100a)$$
$$f_t + 2ff_x + 2g_x - f_{xx} = 0. \qquad (2.100b)$$

So, the problem of determining compatibility for the pair (2.96) turns to a very complicated problem, as we now have to solve a coupled nonlinear system of PDEs for f and g. The good news is, we can actually solve this system (see Sect. 3.5.1).

In light of our discussion so far, we have been seeking compatible equations with the diffusion equation that are linear. Must we assume this on the onset? The answer is no, as the next example illustrates.

Determine functions $f(u)$ such that

$$u_t = u_{xx} \quad \text{and} \quad u_t = f(u)u_x^2, \qquad (2.101)$$

are compatible. As we have done previously, we rewrite (2.101) as

$$u_t = f(u)u_x^2, \quad u_{xx} = f(u)u_x^2, \qquad (2.102)$$

and the compatibility of (2.102) requires that $(u_t)_{xx} - (u_{xx})_t = 0$. It follows from the first of (2.102) that

$$\begin{aligned} u_{tx} &= 2fu_xu_{xx} + f'u_x^3 \\ &= 2f^2u_x^3 + f'u_x^3 \\ &= \left(f' + 2f^2\right)u_x^3. \end{aligned} \qquad (2.103)$$

Note that the second of (2.102) has been used. Further,

$$\begin{aligned} u_{txx} &= \left(f'' + 4ff'\right)u_x^4 + 3\left(f' + 2f^2\right)u_x^2u_{xx} \\ &= \left(f'' + 4ff'\right)u_x^4 + 3\left(f' + 2f^2\right)fu_x^4 \\ &= \left(f'' + 7ff' + 6f^3\right)u_x^4. \end{aligned} \qquad (2.104)$$

Similarly,

$$\begin{aligned} u_{xxt} &= 2fu_xu_{xt} + f'u_tu_x^2 \\ &= 2fu_x\left(f' + 2f^2\right)u_x^3 + f'fu_x^4 \\ &= \left(3ff' + 4f^3\right)u_x^4. \end{aligned} \qquad (2.105)$$

Therefore, compatibility gives rise to

$$\left(f'' + 7ff' + 6f^3\right)u_x^4 - \left(3ff' + 4f^3\right)u_x^4 = 0$$

or

$$\left(f'' + 4ff' + 2f^3\right)u_x^4 = 0$$

and since $u_x \neq 0$, then

$$f'' + 4ff' + 2f^3 = 0. \tag{2.106}$$

The solution of (2.106) is rather complicated but it does admit the simple solution $f(u) = \dfrac{1}{u}$ and thus

$$u_t = \frac{u_x^2}{u}, \quad u_t = u_{xx} \tag{2.107}$$

are compatible.

Up to this point, we have been determining compatible equations for the diffusion equation—a linear second order PDE. We would really like to harness the power of this technique to hopefully gain some insight into nonlinear PDEs. The next example illustrates this very nicely.

Example 2.5 Seek compatibility with the nonlinear diffusion equation

$$u_t = \left(u^m u_x\right)_x, \quad m \neq 0 \tag{2.108}$$

and the linear first order PDE

$$u_t + A(t, x)u_x = B(t, x)u. \tag{2.109}$$

Isolating u_t and u_{xx} from (2.108) and (2.109) gives

$$u_t = B(t, x)u - A(t, x)u_x, \quad u_{xx} = \frac{B(t, x)u - A(t, x)u_x}{u^m} - \frac{mu_x^2}{u}, \tag{2.110}$$

and requiring compatibility $(u_{xx})_t = (u_t)_{xx}$ gives (using (2.110) where appropriate)

$$(A_t + 2AA_x - mAB)\frac{u_x}{u^m} + (2(m+1)B_x - A_{xx})u_x$$

$$+B_{xx}u + \frac{mB^2 - B_t - 2BA_x}{u^{m-1}} = 0.$$

Since A and B are independent of u and u_x, we obtain

$$A_t + 2AA_x - mAB = 0, \tag{2.111a}$$
$$A_{xx} - 2(m+1)B_x = 0, \tag{2.111b}$$
$$B_t + 2BA_x - mB^2 = 0, \tag{2.111c}$$
$$B_{xx} = 0. \tag{2.111d}$$

At this point we impose compatibility for A and B, i.e. $(A_{xx})_t = (A_t)_{xx}$ and $(B_{xx})_t = (B_t)_{xx}$. This gives

$$(3m+4)B_x \left((m+1)B - A_x\right) = 0, \tag{2.112a}$$
$$(3m+4)B_x^2 = 0, \tag{2.112b}$$

and from (2.112b) we see two cases emerge:

(i) $m = -4/3$,

(ii) $m \neq -4/3$ (so $B_x = 0$).

Case (i) $m = -4/3$
In this case we solve (2.111b) and (2.111d) for A and B giving

$$A = -\frac{1}{3}a(t)x^2 + c(t)x + d(t), \quad B = a(t)x + b(t), \tag{2.113}$$

where $a - d$ are arbitrary functions of integration. Further, substitution into the remaining equations of (2.111) gives

$$-\frac{1}{3}\left(\dot{a} + 2ac + \frac{4}{3}ab\right)x^2 + \left(\dot{c} + 2c^2 + \frac{4}{3}bc\right)x + \dot{d} + 2cd + \frac{4}{3}bd = 0, \tag{2.114a}$$
$$\left(\dot{a} + 2ac + \frac{4}{3}ab\right)x + \dot{b} + 2bc + \frac{4}{3}b^2 = 0, \tag{2.114b}$$

and since $a - d$ are independent of x, this leads to

$$\dot{a} + 2ac + \frac{4}{3}ab = 0, \quad \dot{b} + 2bc + \frac{4}{3}b^2 = 0, \tag{2.115a}$$
$$\dot{c} + 2c^2 + \frac{4}{3}bc = 0, \quad \dot{d} + 2cd + \frac{4}{3}cd = 0. \tag{2.115b}$$

We will assume that $a \neq 0$ as this is contained in case (ii). Eliminating the nonlinear terms in (2.115) gives

$$b\dot{a} - a\dot{b} = 0, \quad c\dot{a} - a\dot{c} = 0, \quad d\dot{a} - a\dot{d} = 0, \tag{2.116}$$

and are easily solved giving

$$b = k_1 a, \quad c = k_2 a, \quad d = k_3 a, \tag{2.117}$$

where k_1, k_2 and k_3 are arbitrary constants. With these assignments, (2.115a) becomes

$$\dot{a} + \left(2k_2 + \frac{4}{3}k_1\right)a^2 = 0, \tag{2.118}$$

and is easily solved giving

$$a = \frac{1}{\left(2k_2 + \frac{4}{3}k_1\right)t + k_4}, \tag{2.119}$$

where k_4 is an additional constant. Thus, from (2.117) and (2.119) we obtain the forms for a, b, c and d, and through (2.113), the forms of A and B. Thus, (2.109) is compatible with equations of the form

$$u_t + \frac{-\frac{1}{3}x^2 + k_2x + k_3}{\left(2k_2 + \frac{4}{3}k_1\right)t + k_4} u_x = \frac{x + k_2}{\left(2k_2 + \frac{4}{3}k_1\right)t + k_4} u. \tag{2.120}$$

Case (ii) $m \neq -4/3$

In this case, we have $B_x = 0$. We solve (2.111b) and (2.111d) giving

$$A = c(t)x + d(t), \quad B = b(t), \tag{2.121}$$

where $b, c,$ and d are arbitrary functions of integration. Further, substitution into the remaining equations of (2.111) gives

$$\left(\dot{c} + 2c^2 - mbc\right)x + \dot{d} + 2cd - mbd = 0, \tag{2.122a}$$

$$\dot{b} + 2bc - mb^2 = 0, \tag{2.122b}$$

and since $b, c,$ and d are independent of x, this leads to

$$\dot{b} + 2bc - mb^2 = 0, \quad \dot{c} + 2c^2 - mbc = 0, \quad \dot{d} + 2cd - mbc = 0. \tag{2.123a}$$

These can be solved as in the previous case. Their solution is:
If $b = 0$ then

$$c = \frac{c_0}{2c_0t + k}, \quad d = \frac{d_0}{2c_0t + k}, \tag{2.124}$$

giving (2.109) as

$$u_t + \frac{2c_0x + d_0}{2c_0t + k} u_x = 0. \tag{2.125}$$

If $b \neq 0$ then

$$b = \frac{1}{(2k_1 - m)t + k}, \quad c = \frac{c_0}{(2k_1 - m)t + k}, \quad d = \frac{d_0}{(2k_1 - m)t + k}, \tag{2.126}$$

where b_0, c_0, d_0 and k are all arbitrary constants. The result from (2.109) is

$$u_t + \frac{c_0 x + d_0}{(2k_1 - m)b_0 t + k} u_x = \frac{b_0}{(2k_1 - m)b_0 t + k} u. \tag{2.127}$$

We now would like to use these results to find exact solutions of the original PDE (2.108). In the case of $m = -4/3$, we use (2.120). We will consider the case where $k_1 = k_2 = k_3 = 0$ and $k_4 = 1$. So (2.120) becomes

$$u_t - \frac{1}{3} x^2 u_x = x\, u. \tag{2.128}$$

Solving (2.128) gives

$$u = x^{-3} F\left(\frac{3}{x} - t\right), \tag{2.129}$$

and substituting into (2.108) gives

$$-F' = 9 \left(\frac{F'}{F^{4/3}}\right)'. \tag{2.130}$$

Integrating once gives

$$-F = 9 \frac{F'}{F^{4/3}} + c_1. \tag{2.131}$$

Even though (2.131) can be solved leading to an implicit solution, we will suppress the constant c_1. This leads to the exact solution

$$F = \left(\frac{27}{4} \frac{1}{\lambda + \lambda_0}\right)^{3/4}, \tag{2.132}$$

and through (2.129) gives

$$u = x^{-3} \left(\frac{27}{4} \frac{1}{\dfrac{3}{x} - t + \lambda_0}\right)^{3/4}. \tag{2.133}$$

Next, we will use (2.125). We will consider the case where $k = d_0 = 0$. So (2.125) becomes

$$u_t + \frac{x}{2t} u_x = 0. \tag{2.134}$$

Solving gives

$$u = F\left(\frac{x}{\sqrt{t}}\right), \tag{2.135}$$

and substituting into (2.108) gives

$$-\frac{1}{2} \lambda F' = F^m F'' + m F^{m-1} F'^2, \quad \lambda = x/\sqrt{t}. \tag{2.136}$$

Now any solution of this ODE will give rise to an exact solution of the original PDE (2.108).

Example 2.6 Seek compatibility of Burgers' equation

$$u_t + uu_x = u_{xx} \tag{2.137}$$

and the quasilinear first order PDE

$$u_t + X(t, x, u)u_x = U(t, x, u). \tag{2.138}$$

Isolating u_t and u_{xx} from (2.137) and (2.138) gives

$$u_t = U(t, x, u) - X(t, x, u)u_x, \quad u_{xx} = U(t, x, u) + (u - X(t, x, u))u_x, \tag{2.139}$$

and requiring compatibility $(u_{xx})_t = (u_t)_{xx}$ (using (2.139) where appropriate) gives

$$\begin{aligned}
X_{uu}u_x^3 + (2X_{xu} - U_{uu} + 2uX_u - 2XXu)\,u_x^2 \\
(X_{xx} - 2U_{xu} - X_t + uX_x - 2XX_x + 2UX_u + U)\,u_x \\
+U_t + 2UX_x + uU_x - U_{xx} = 0.
\end{aligned} \tag{2.140}$$

Since X and U are independent of u_x, for (2.140) to be identically satisfied, the coefficients of u_x^m, $m = 0, 1, 2$ and 3 must be zero. This gives

$$X_{uu} = 0, \tag{2.141a}$$
$$2X_{xu} - U_{uu} + 2uX_u - 2XXu = 0, \tag{2.141b}$$
$$X_{xx} - 2U_{xu} - X_t + uX_x - 2XX_x + 2UX_u + U = 0, \tag{2.141c}$$
$$U_t + 2UX_x + uU_x - U_{xx} = 0. \tag{2.141d}$$

We can easily solve the first two equations in (2.141). This gives rise to

$$X = Ku + A, \tag{2.142a}$$
$$U = \frac{1}{3}K(1 - K)u^3 + (K_x - AK)^2 + Bu + C, \tag{2.142b}$$

where K, A, B and C are arbitrary functions of t and x. Substituting (2.142) into (2.141c) and (2.141d) and isolating coefficients with respect to u gives

$$C_t + 2CA_x - C_{xx} = 0,$$

$$(2.143a)$$

$$B_t + C_x + 2BA_x + 2CK_x - B_{xx} = 0,$$

$$(2.143b)$$

$$A(K_{xx} - K_t) + K_{tx} - K_{xxx} + B_x + K(A_{xx} - 2AA_x - A_t) + (4A_x + 2B)K_x = 0,$$

$$(2.143c)$$

$$\frac{1}{3}(1 - 2K)K_t - (2K + 1)AK_x + \frac{8}{3}K_x^2 + \frac{2}{3}(K + 1)K_{xx} - \frac{1}{3}(2K^2 + K)A_x = 0,$$

$$(2.143d)$$

$$-\frac{1}{3}(2K^2 - 1)K_x = 0,$$

$$(2.143e)$$

and

$$A_{xx} - 2AA_x + C - A_t + 2KC - 2B_x = 0, \qquad (2.144a)$$

$$-K_t - 3K_{xx} + 2KB + B + 2KA_x + 2AK_x + A_x = 0, \qquad (2.144b)$$

$$K(4K_x - 2KA - A) = 0, \qquad (2.144c)$$

$$-\frac{1}{3}K(2K + 1)(K - 1) = 0. \qquad (2.144d)$$

The systems (2.143) and (2.144) represent an overdetermined system of equations to determine the form of A, B, C, and K. From (2.144d) we see that three cases emerge:

$$(i) \quad K = 0,$$
$$(ii) \quad K = 1,$$
$$(iii) \quad K = -1/2.$$

Each will be considered separately.

Case (i) $K = 0$

In this case, the systems (2.143) and (2.144) become

$$C_t + 2CA_x - C_{xx} = 0, \qquad (2.145a)$$
$$B_t + C_x + 2BA_x - B_{xx} = 0, \qquad (2.145b)$$
$$B_x = 0, \qquad (2.145c)$$

and

$$A_{xx} - A_t - 2B_x - 2AA_x + C = 0, \tag{2.146a}$$

$$B + A_x = 0. \tag{2.146b}$$

From (2.145c), (2.146a), and (2.146b) we find

$$A = a(t)x + b(t), \quad B = -a(t), \quad C = \left(\dot{a} + 2a^2\right)x + \dot{b} + 2ab, \tag{2.147}$$

where a and b are arbitrary functions of t. Here we find that (2.145b) is automatically satisfied while (2.145a) becomes

$$\left(\ddot{a} + 6a\dot{a} + 4a^3\right)x + \ddot{b} + 2b\dot{a} + 4a^2b = 0, \tag{2.148}$$

which leads to

$$\ddot{a} + 6a\dot{a} + 4a^3 = 0, \tag{2.149a}$$

$$\ddot{b} + 2b\dot{a} + 4a^2b = 0. \tag{2.149b}$$

If we let

$$a = \frac{1}{2}\frac{\dot{c}}{c}, \quad b = \frac{1}{2}\frac{\dot{d}}{c}, \tag{2.150}$$

then (2.149) becomes

$$\frac{\dddot{c}}{c} = 0, \quad \frac{\dddot{d}}{c} = 0, \tag{2.151}$$

which readily integrates, giving

$$c = c_2 t^2 + c_1 t + c_0, \quad d = c_3 t^2 + c_4 t + c_5, \tag{2.152}$$

where $c_0 - c_5$ are arbitrary constants leading to a and b as

$$a = \frac{1}{2}\frac{2c_2 t + c_1}{c_2 t^2 + c_1 t + c_0}, \quad b = \frac{1}{2}\frac{2c_3 t + c_4}{c_2 t^2 + c_1 t + c_0}. \tag{2.153}$$

This gives compatible equations to Burgers' equation in the form

$$u_t + \frac{1}{2}\left(\frac{(2c_2 t + c_1)x + 2c_3 t + c_4}{c_2 t^2 + c_1 t + c_0}\right)u_x = -\frac{1}{2}\frac{2c_2 t + c_1}{c_2 t^2 + c_1 t + c_0}u + \frac{c_2 x + c_3}{c_2 t^2 + c_1 t + c_0}. \tag{2.154}$$

Case (ii) $K = 1$
In this case, the systems (2.143) and (2.144) become

$$C_t + 2CA_x - C_{xx} = 0, \tag{2.155a}$$
$$B_t + C_x + 2BA_x - B_{xx} = 0, \tag{2.155b}$$
$$-A_t - 2AA_x + B_x + A_{xx} = 0, \tag{2.155c}$$
$$-A_x = 0 \tag{2.155d}$$

and

$$A_{xx} - A_t - 2B_x - 2AA_x + 3C = 0, \tag{2.156a}$$
$$3B + 3A_x = 0, \tag{2.156b}$$
$$-3A = 0. \tag{2.156c}$$

From (2.156) we readily see that $A = B = C = 0$ and that (2.155) is automatically satisfied. From (2.142) we obtain $X = u$ and $U = 0$. From (2.137) and (2.138) we obtain

$$u_t + uu_x = 0, \quad u_{xx} = 0, \tag{2.157}$$

which has the common solution

$$u = \frac{c_1 x + c_3}{c_1 t + c_2}, \tag{2.158}$$

where $c_1 - c_3$ are arbitrary constants.

Case (iii) $K = -1/2$

In this case, the systems (2.143) and (2.144) become

$$C_t + 2CA_x - C_{xx} = 0,$$
$$B_t + C_x + 2BA_x - B_{xx} = 0, \tag{2.159}$$
$$A_t + 2AA_x + 2B_x - A_{xx} = 0.$$

It is interesting to note that in this case, we are required to solve the nonlinear system of PDEs (2.159). A similar result was obtained earlier when we sought compatibility with the diffusion equation (Eq. (2.100)). Maybe these two equations, the heat equation and Burgers equation, are connected.

Example 2.7 Seek compatibility with the linear diffusion equation with a source term

$$u_t = u_{xx} + Q(u), \quad Q'' \neq 0, \tag{2.160}$$

and general quasilinear first order PDE

$$u_t + X(t, x, u)u_x = U(t, x, u). \tag{2.161}$$

Isolating u_t and u_{xx} from (2.160) and (2.161) gives

$$u_t = U(t, x, u) - X(t, x, u)u_x \quad \text{and} \quad u_{xx} = U(t, x, u) - X(t, x, u)u_x - Q(u),$$
$$(2.162)$$

and requiring compatibility $(u_{xx})_t = (u_t)_{xx}$ (using (2.162) where appropriate) gives

$$X_{uu}u_x^3 + (2X_{xu} - U_{uu} - 2XX_u)\,u_x^2$$
$$+ (X_{xx} - 2U_{xu} + 2UX_u - 3QX_u - 2XX_x - X_t)\,u_x \qquad (2.163)$$
$$+ U_t - U_{xx} + 2UX_x + QU_u - 2QX_x - UQ' = 0.$$

Since X and U are independent of u_x we obtain

$$X_{uu} = 0, \qquad (2.164\text{a})$$
$$2X_{xu} - U_{uu} - 2XX_u = 0, \qquad (2.164\text{b})$$
$$X_{xx} - 2U_{xu} + 2UX_u - 3QX_u - 2XX_x - X_t = 0, \qquad (2.164\text{c})$$
$$U_t - U_{xx} + 2UX_x + QU_u - 2QX_x - UQ' = 0. \qquad (2.164\text{d})$$

Integrating (2.164a) and (2.164b) gives

$$X = Pu + A, \quad U = -\frac{1}{3}P^2u^3 + (P_x - AP)\,u^2 + Bu + C, \qquad (2.165)$$

where A, B, C and P are arbitrary functions of t and x. Substituting further into (2.164c) gives

$$3PQ + \frac{2}{3}P^3u^3 - 2P\,(2P_x - AP)\,u^2$$
$$+ (P_t + 3P_{xx} - 2PA_x - 2AP_x - 2BP)\,u \qquad (2.166)$$
$$A_t - A_{xx} + 2AA_x + 2B_x - 2CP = 0.$$

From (2.166) we find that if $P \neq 0$ then we have a preliminary from of Q. Thus, two cases emerge:

$$(i) \quad P \neq 0,$$
$$(ii) \quad P = 0.$$

Case 1 $P \neq 0$ Dividing (2.166) by $3P$ and isolating Q gives

$$Q = -\frac{2}{9}P^2u^3 + \frac{2}{3}(2P_x - AP)\,u^2 - \left(\frac{P_t + 3P_{xx} - 2PA_x - 2AP_x - 2BP}{3P}\right)u$$
$$-\frac{A_t - A_{xx} + 2AA_x + 2B_x - 2CP}{3P}. \qquad (2.167)$$

As $Q = Q(u)$ only, from (2.167) we deduce that $P = p$, $A = a$, $B = b$ and $C = c$, all constant. Thus, Q has the form

$$Q = -\frac{2}{9}p^2u^3 - \frac{2}{3}apu^2 + \frac{2}{3}bu + \frac{2}{3}c. \tag{2.168}$$

With this choice of Q we find the remaining equation in (2.164) is automatically satisfied. Thus, we find

$$X = pu + a, \quad U = -\frac{1}{3}p^2u^3 - pau^2 + bu + c, \tag{2.169}$$

and the following two are compatible

$$u_t = u_{xx} - \frac{2}{9}p^2u^3 - \frac{2}{3}apu^2 + \frac{2}{3}bu + \frac{2}{3}c, \tag{2.170a}$$

$$u_t + (pu + a)u_x = -\frac{1}{3}p^2u^3 - pau^2 + bu + c. \tag{2.170b}$$

Case 2 $P = 0$ In this case, (2.165) reduces to

$$X = A, \quad U = Bu + C, \tag{2.171}$$

while (2.164c) and (2.164d) become

$$A_t + 2AA_x + 2B_x - A_{xx} = 0, \tag{2.172}$$

and

$$(Bu + C)Q' + (2A_x - B)Q = (B_t + 2BA_x - B_{xx})u + C_t + 2CA_x - C_{xx}. \tag{2.173}$$

Since (2.173) should only be an ODE for Q, four cases arise:

$$
\begin{array}{lll}
\text{(i)} & B = 0, & C = 0, \\
\text{(ii)} & B = 0, & C \neq 0, \\
\text{(iii)} & B \neq 0, & C = 0, \\
\text{(iv)} & B \neq 0, & C \neq 0.
\end{array}
$$

Each will be considered separately.

Subcase (i) $B = 0$, $C = 0$
In this case, (2.172) and (2.173) become

$$A_t + 2AA_x - A_{xx} = 0, \quad 2A_x Q(u) = 0, \tag{2.174}$$

from which we deduce that $A = a$, a constant, since $Q \neq 0$. Thus, for arbitrary $Q(u)$, the following are compatible

$$u_t = u_{xx} + Q(u), \quad u_t + au_x = 0. \tag{2.175}$$

Subcase (ii) $B = 0, \ C \neq 0$
In this case, Eq. (2.173) becomes

$$CQ' + 2A_x Q = C_t + 2CA_x - C_{xx}, \tag{2.176}$$

and dividing through by C gives

$$Q'(u) + \frac{2A_x}{C} Q(u) = \frac{C_t + 2CA_x - C_{xx}}{C}. \tag{2.177}$$

Since (2.177) should be only an equation involving u, then

$$\frac{2A_x}{C} = -m, \tag{2.178a}$$

$$\frac{C_t + 2CA_x - C_{xx}}{C} = k_1, \tag{2.178b}$$

where m and k_1 are arbitrary constants. With these assignments, (2.177) becomes

$$Q'(u) - mQ(u) = k_1. \tag{2.179}$$

We note that $m \neq 0$, as $m = 0$ gives $Q'' = 0$ which is inadmissible. From (2.178a), we solve for C giving

$$C = -\frac{2A_x}{m}, \tag{2.180}$$

and with this choice (2.178b) becomes

$$A_{tx} - A_{xxx} + 2A_x^2 - k_1 A_x = 0. \tag{2.181}$$

Within this subcase (2.172) becomes

$$A_t + 2AA_x - A_{xx} = 0. \tag{2.182}$$

It is left as an exercise to the reader to show that if (2.181) and (2.182) are consistent, then $k_1 A_x = 0$; and since $A_x = 0$ gives $C = 0$, then we require that $k_1 = 0$. We can readily solve (2.181) and (2.182) for A giving

$$A = \frac{c_1 x + c_2}{2c_1 t + c_0}, \tag{2.183}$$

where $c_0 - c_2$ are arbitrary constants while solving (2.179) gives

$$Q(u) = k_2 e^{mu}, \tag{2.184}$$

where k_2 is an additional arbitrary constant. Thus, the following are compatible

$$u_t = u_{xx} + k_2 e^{mu}, \quad u_t + \frac{c_1 x + c_2}{2c_1 t + c_0} u_x = -\frac{2c_1}{m(2c_1 t + c_0)}. \tag{2.185}$$

Subcase (iii) $B \neq 0, \ C = 0$
In this case, Eq. (2.173) becomes

$$Bu Q'(u) + (2A_x - B) Q(u) = (B_t + 2BA_x - B_{xx}) u, \tag{2.186}$$

and dividing through by B gives

$$u Q'(u) + \frac{2A_x - B}{B} Q(u) = \frac{B_t + 2BA_x - B_{xx}}{B} u. \tag{2.187}$$

Since (2.187) should be only an equation involving u, then

$$\frac{2A_x - B}{B} = -m, \tag{2.188a}$$

$$\frac{B_t + 2BA_x - B_{xx}}{B} = k_1, \tag{2.188b}$$

where m and k_1 are arbitrary constants and (2.187) becomes

$$u Q'(u) - m Q(u) = k_1 u. \tag{2.189}$$

We can solve (2.188a) for B explicitly provided that $m \neq 1$. As this is a special case, we consider it first.

Special Case: $m = 1$
If $m = 1$, then from (2.189)

$$u Q'(u) - Q(u) = k_1 u, \tag{2.190}$$

whose solution is

$$Q(u) = (k_1 \ln u + k_2) u, \tag{2.191}$$

where k_2 is a constant of integration. For this case (2.188a) becomes

$$A_x = 0, \tag{2.192}$$

which must be solved in conjunction with (2.172) and (2.188b). These are easily solved giving

$$A = c_1 e^{k_1 t} + c_2, \quad B = c_3 e^{k_1 t} - \frac{1}{2} c_1 k_1 e^{k_1 t} x, \tag{2.193}$$

where c_1, c_2 and c_3 are arbitrary constants. Thus, the following are compatible

$$u_t = u_{xx} + (k_1 \ln u + k_2) u, \quad u_t + \left(c_1 e^{k_1 t} + c_2 \right) u_x = \left(c_3 e^{k_1 t} - \frac{1}{2} c_1 k_1 e^{k_1 t} x \right) u.$$

$$(2.194)$$

In the case where $m \neq 1$, we solve (2.188a) for B giving

$$B = \frac{2 A_x}{1 - m}. \tag{2.195}$$

Thus, (2.172) and (2.188b) become

$$A_t + 2 A A_x + \frac{3 + m}{1 - m} A_{xx} = 0, \quad A_{tx} + 2 A_x^2 - A_{xxx} = k_1 A_x. \tag{2.196}$$

These are compatible provided that either $k_1 = 0$ or $m = 3$. Of course there are other choices but we require that $B \neq 0$ which eliminate those like $A = 0$ and $A_x = 0$. The solution for A in each case is

$$A = \frac{c_1 x + c_3}{2 c_1 t + c_0} \ (k_1 = 0), \quad A = -3 \frac{S'}{S} \ (m = 3), \tag{2.197}$$

where $S = S(x)$ satisfies $S'' + \frac{k_1}{4} S = 0$. For each case B is given through (2.195) and Q is obtained through (2.189). Our final results:

$$u_t = u_{xx} + k_2 u^m, \quad u_t + \frac{c_1 x + c_3}{2 c_1 t + c_0} u_x = \frac{2 c_1 u}{(1 - m)(2 c_1 t + c_0)}, \tag{2.198}$$

and

$$u_t = u_{xx} + k_2 u^3 - \frac{1}{2} k_1 u, \quad u_t - 3 \frac{S'}{S} u_x = 3 \left(\frac{S'}{S} \right)' u, \tag{2.199}$$

are compatible.

Case (iv) $B \neq 0, \ C \neq 0$
In this case, dividing (2.189) by B gives

$$\left(u + \frac{C}{B} \right) Q' + \frac{(2 A_x - B)}{B} Q = \frac{(B_t + 2 B A_x - B_{xx})}{B} u + \frac{C_t + 2 C A_x - C_{xx}}{B}. \tag{2.200}$$

Since (2.200) should be only an equation involving u, then

$$\frac{C}{B} = a, \tag{2.201a}$$

$$\frac{(2A_x - B)}{B} = -m, \tag{2.201b}$$

$$\frac{(B_t + 2BA_x - B_{xx})}{B} = k_1, \tag{2.201c}$$

$$\frac{C_t + 2CA_x - C_{xx}}{B} = k_2, \tag{2.201d}$$

where a, m, k_1 and k_2 are arbitrary constants. From (2.201) we deduce $k_2 = ak_1$ and from (2.200) that Q satisfies

$$(u + a)Q'(u) - mQ(u) = k_1(u + a), \tag{2.202}$$

which is (2.189) with a translation in the argument u. Since we can translate u in the original PDE without loss of generality, we can set $a = 0$ without loss of generality, and thus we are led back to case (iii).

Exact Solutions

One of the main goals of deriving compatible equations is to use them to construct exact solutions of a given PDE. As cubic source terms appear to be special we will focus on these and consider PDEs of the form

$$u_t = u_{xx} + q_1 u^3 + q_2 u \tag{2.203}$$

where q_1 and q_2 are constant. This PDE is know as the Newel-Whitehead-Segel equation and was introduced by Newel and Whitehead [4] and Segel [5] to model various phenomena in fluid mechanics.

As cubic source terms arose in two places in our study, we consider each separately. We first consider (2.170). Here we will set $a = 0$, $b = -3k$ and $c = 0$, where k is constant, and for convenience we choose $p = 3$. In this case (2.170) becomes

$$u_t + 3uu_x = -3u^3 - 3ku, \tag{2.204a}$$

$$u_t = u_{xx} - 2u^3 - 2ku. \tag{2.204b}$$

Eliminating u_t in (2.204) gives

$$u_{xx} + 3uu_x + u^3 + ku = 0 \tag{2.205}$$

which involves the second member of the Riccati hierarchy. Under the substitution $u = S_x/S$, (2.205) becomes

$$S_{xxx} + kS_x = 0. \tag{2.206}$$

The solution of (2.206) differs depending on the sign of k and is given as:

$$S(t) = s_2(t)e^{\omega x} + s_1(t)e^{-\omega x} + s_0(t) \quad (k = -\omega^2), \tag{2.207a}$$

$$S(t) = s_2(t)x^2 + s_1(t)x + s_0(t) \quad (k = 0), \tag{2.207b}$$

$$S(t) = s_2(t)\sin(\omega x) + s_1(t)\cos(\omega x) + s_0(t) \quad (k = \omega^2), \tag{2.207c}$$

where s_1, s_2 and s_3 are arbitrary functions of integration. Substitution of any of these into (2.204a) leads to the following ODEs for s_1, s_2 and s_3 .

$$s_1 s_2' - s_2 s_1' = 0, \tag{2.208a}$$

$$s_2 s_0' - s_0 s_2' - 3k s_0 s_2 = 0, \tag{2.208b}$$

for $k \neq 0$ and

$$s_1 s_2' - s_2 s_1' = 0, \tag{2.209a}$$

$$s_2 s_0' - s_0 s_2' - 6s_2^2 = 0, \tag{2.209b}$$

for $k = 0$. These are easily solved, giving

$$S(t) = \left(c_1 e^{\omega x} + c_2 e^{-\omega x} + c_3 e^{-3\omega^2 t}\right) s(t) \quad (k = -\omega^2), \tag{2.210a}$$

$$S(t) = \left(c_1(x^2 + 6t) + c_2 x + c_3\right) s(t) \quad (k = 0), \tag{2.210b}$$

$$S(t) = \left(c_1 \sin(\omega x) + c_2 \cos(\omega x) + c_3 e^{3\omega^2 t}\right) s(t) \quad (k = \omega^2), \tag{2.210c}$$

where $c_1 - c_3$ are constant and via $u = S_x/S$ gives the exact solutions

$$u = \frac{\omega\left(c_1 e^{\omega x} - c_2 e^{-\omega x}\right)}{c_1 e^{\omega x} + c_2 e^{-\omega x} + c_3 e^{-3\omega^2 t}} \quad (k = -\omega^2),$$

$$u = \frac{2c_1 x + c_2}{c_1(x^2 + 6t) + c_2 x + c_3} \quad (k = 0), \tag{2.211}$$

$$u = \frac{\omega\left(c_1 \cos(\omega x) - c_2 \sin(\omega x)\right)}{c_1 \cos(\omega x) + c_2 \sin(\omega x) + c_3 e^{3\omega^2 t}} \quad (k = \omega^2).$$

The second place we saw cubic source terms was in (2.199). We will set $k_1 = 4k$ and $k_2 = q$ for convenience, so (2.199) becomes

$$u_t - 3\frac{S'}{S}u_x = 3\left(\frac{S'}{S}\right)' u, \tag{2.212a}$$

$$u_t = u_{xx} + qu^3 - 2ku, \tag{2.212b}$$

where S satisfies $S'' + kS = 0$. We will distinguish between two different cases:

(i) $k = 0$,

(ii) $k \neq 0$.

In the case of $k = 0$, then $S = c_1 x + c_2$. However, we can set $c_1 = 1$ and $c_2 = 0$ without loss of generality. In this case (2.212a) becomes

$$u_t - \frac{3}{x} u_x = -\frac{3}{x^2} u,$$ (2.213)

which is solved giving

$$u = x F(x^2 + 6t),$$ (2.214)

where F is an arbitrary function of its argument. Substitution of (2.214) into (2.212b) (with $k = 0$) gives

$$F'' + \frac{q}{4} F^3 = 0,$$ (2.215)

where $F = F(\lambda)$, $\lambda = x^2 + 6t$. The solution of (2.215) can be expressed in terms of the Jacobi elliptic functions so

$$F = \frac{2}{\sqrt{q}} cn(\lambda, \tfrac{1}{\sqrt{2}}) \quad \text{and} \quad F = \frac{2}{\sqrt{-q}} nc(\lambda, \tfrac{1}{\sqrt{2}})$$

depending on the sign of q.

In the case of $k \neq 0$, we integrate (2.212a), giving

$$u = \frac{k S(x)}{S'(x)} F\left(3kt - \ln(S')\right),$$ (2.216)

where F is again an arbitrary function of its argument; substitution into (2.212b) gives rise to the ODE

$$F'' + 3F' + 2F + q F^3 = 0,$$ (2.217)

where $F = F(\lambda)$, $\lambda = 3kt - \ln(S')$. If we introduce the change of variables

$$\lambda = -\ln \xi, \quad F = \xi G,$$ (2.218)

where $G = G(\xi)$, then (2.217) becomes

$$G'' + q G^3 = 0,$$ (2.219)

which is essentially (2.215). If we impose the change of variables (2.218) on the solution (2.216), then we obtain

$$u = k S(x) e^{-3kt} G\left(S'(x) e^{-3kt}\right),$$ (2.220)

and substitution of (2.220) into (2.212b) gives (2.219) directly! Furthermore, as $kS = -S''$, we can write (2.220)

$$u = -S''(x) e^{-3kt} G\left(S'(x) e^{-3kt}\right)$$ (2.221)

or

$$u = \bar{S}'(x) e^{-3kt} G\left(\bar{S}(x) e^{-3kt}\right),$$ (2.222)

where $\bar{S}$ satisfies the same equation as $-S$. Thus, we again obtain solutions in terms of Jacobi Elliptic functions $u = \dfrac{\bar{S}'(x)e^{-3kt}}{\sqrt{q}} \, cn(\xi, \frac{1}{\sqrt{2}})$ and $u = \dfrac{\bar{S}'(x)e^{-3kt}}{\sqrt{-q}} \, nc(\xi, \frac{1}{\sqrt{2}})$, $\xi = \bar{S}(x)e^{-3kt}$ depending on the sign of q. We note that these exact solutions appear in *EqWorld* [6].

2.3   Compatibility in $(2+1)$ Dimensions

Up until now, we have considered compatibility of PDEs in $(1+1)$ dimensions. We now extend this idea and consider PDEs in $(2+1)$ dimensions. In this section we consider the compatibility between the $(2+1)$ dimensional reaction $-$ diffusion equation

$$u_t = u_{xx} + u_{yy} + Q(u, u_x, u_y),\qquad(2.223)$$

and the first order partial differential equation

$$u_t = F\left(t, x, y, u, u_x, u_y\right).\qquad(2.224)$$

This section is based on the work of Arrigo and Suazo [3]. We will assume that F in (2.224) is nonlinear in the first derivatives u_x and u_y. The case where F is linear in the first derivatives u_x and u_y, is left as an exercise to the reader.

Compatibility between (2.223) and (2.224) gives rise to the compatibility equation constraints

$$F_{pp} + F_{qq} = 0,\qquad(2.225a)$$
$$F_{xp} - F_{yq} + pF_{up} - qF_{uq} + (F - Q)F_{pp} = 0,\qquad(2.225b)$$
$$F_{xq} + F_{yp} + qF_{up} + pF_{uq} + (F - Q)F_{pq} = 0,\qquad(2.225c)$$
$$-F_t + F_{xx} + F_{yy} + 2pF_{xu} + 2qF_{yu} + 2(F - Q)F_{yq} +$$
$$\left(p^2 + q^2\right)F_{uu} + 2q(F - Q)F_{uq} + (F - Q)^2 F_{qq} +$$
$$Q_pF_x + Q_qF_y + \left(pQ_p + qQ_q - Q\right)F_u - pQ_uF_p - qQ_uF_q + FQ_u = 0.\qquad(2.225d)$$

Eliminating the x and y derivatives in (2.225b) and (2.225c) by (i) cross differentiation and (ii) imposing (2.225a) gives

$$2F_{up} + (F_p - Q_p)F_{pp} + (F_q - Q_q)F_{pq} = 0,\qquad(2.226a)$$
$$2F_{uq} + (F_p - Q_p)F_{pq} + (F_q - Q_q)F_{qq} = 0.\qquad(2.226b)$$

Further, eliminating F_{up} and F_{uq} by again (i) cross differentiation and (ii) imposing (2.225a) gives rise to

$$(2F_{pp} - Q_{pp} + Q_{qq})F_{pp} + 2(F_{pq} - Q_{pq})F_{pq} = 0, \qquad (2.227a)$$

$$(Q_{pp} - Q_{qq})F_{pq} + 2Q_{pq}F_{qq} = 0. \qquad (2.227b)$$

Solving (2.225a), (2.227a) and (2.227b) for F_{pp}, F_{pq} and F_{qq} gives rise to two cases:

$$(i) \quad F_{pp} = F_{pq} = F_{qq} = 0, \qquad (2.228a)$$

$$(ii) \quad F_{pp} = \frac{1}{2}(Q_{pp} - Q_{qq}), \quad F_{pq} = Q_{pq}, \quad F_{qq} = \frac{1}{2}(Q_{qq} - Q_{pp}). \qquad (2.228b)$$

As we are primarily interested in compatible equations that are more general than quasilinear, we omit the first case. If we require that the three equations in (2.228b) be compatible, then to within equivalence transformations of the original equation, Q satisfies

$$Q_{pp} + Q_{qq} = 0. \qquad (2.229)$$

Using (2.229), we find that (2.228b) becomes

$$F_{pp} = Q_{pp}, \quad F_{pq} = Q_{pq}, \quad F_{qq} = Q_{qq}, \qquad (2.230)$$

from which we find that

$$F = Q(u, p, q) + X(t, x, y, u)p + Y(t, x, y, u)q + U(t, x, y, u), \qquad (2.231)$$

where X, Y and U are arbitrary functions. Substituting (2.231) into (2.226a) and (2.226b) gives

$$2Q_{up} + XQ_{pp} + YQ_{pq} + 2X_u = 0, \qquad (2.232a)$$

$$2Q_{uq} + XQ_{pq} + YQ_{qq} + 2Y_u = 0, \qquad (2.232b)$$

while (2.225b) and (2.225c) become (using (2.229) and (2.232))

$$(Xp + Yq + 2U)\,Q_{pp} + (Xq - Yp)\,Q_{pq} + 2\left(X_x - Y_y\right) = 0, \qquad (2.233a)$$

$$(Xq - Yp)\,Q_{pp} - (Xp + Yq + 2U)\,Q_{pq} - 2\left(X_y + Y_x\right) = 0. \qquad (2.233b)$$

If we differentiate (2.232a) and (2.232b) with respect to x and y, we obtain

$$X_x Q_{pp} + Y_x Q_{pq} + 2X_{xu} = 0, \quad X_x Q_{pq} + Y_x Q_{qq} + 2Y_{xu} = 0, \qquad (2.234a)$$

$$X_y Q_{pp} + Y_y Q_{pq} + 2X_{yu} = 0, \quad X_y Q_{pq} + Y_y Q_{qq} + 2Y_{yu} = 0. \qquad (2.234b)$$

If $X_x^2 + Y_x^2 \neq 0$, then solving (2.229) and (2.234a) for Q_{pp}, Q_{pq} and Q_{qq} gives

$$Q_{pp} = -Q_{qq} = \frac{2\left(Y_x Y_{xu} - X_x X_{xu}\right)}{X_x^2 + Y_x^2}, \quad Q_{pq} = -\frac{2\left(X_x Y_{xu} + Y_x X_{xu}\right)}{X_x^2 + Y_x^2}.$$

If $X_y^2 + Y_y^2 \neq 0$, then solving (2.229) and (2.234b) for Q_{pp}, Q_{pq} and Q_{qq} gives

$$Q_{pp} = -Q_{qq} = \frac{2\left(Y_y Y_{yu} - X_y X_{yu}\right)}{X_y^2 + Y_y^2}, \quad Q_{pq} = -\frac{2\left(X_y Y_{yu} + Y_y X_{yu}\right)}{X_y^2 + Y_y^2}.$$

In any case, this shows that Q_{pp}, Q_{pq} and Q_{qq} are, at most, functions of u only. Thus, if we let

$$Q_{pp} = -Q_{qq} = 2g_1(u), \quad Q_{pq} = g_2(u),$$

for arbitrary functions g_1 and g_2, then Q has the form

$$Q = g_1(u)\left(p^2 - q^2\right) + g_2(u)pq + g_3(u)p + g_4(u)q + g_5(u), \tag{2.235}$$

where $g_3 - g_5$ are further arbitrary functions. Substituting (2.235) into (2.233) gives

$$2\left(Xp + Yq + 2U\right)g_1 + (Xq - Yp)g_2 + 2\left(X_x - Y_y\right) = 0,$$
$$2\left(Xq - Yp\right)g_1 - (Xp + Yq + 2U)g_2 - 2\left(X_y + Y_x\right) = 0. \tag{2.236}$$

Since both equations in (2.236) must be satisfied for all p and q, this requires that each coefficient of p and q must vanish. This leads to

$$2g_1 X - g_2 Y = 0,$$
$$g_2 X + 2g_1 Y = 0, \tag{2.237}$$

and

$$2g_1 U + X_x - Y_y = 0,$$
$$g_2 U + X_y + Y_x = 0. \tag{2.238}$$

From (2.237) we see that either $g_1 = g_2 = 0$ or $X = Y = 0$. If $g_1 = g_2 = 0$, then Q is quasilinear giving that F is quasilinear, which violates our non-quasilinearity condition. If $X = Y = 0$, we are led to a contradiction, as we imposed $X_x^2 + Y_x^2 \neq 0$ or $X_y^2 + Y_y^2 \neq 0$. Thus, it follows that

$$X_x^2 + Y_x^2 = 0, \quad X_y^2 + Y_y^2 = 0,$$

or

$$X_x = 0, \quad X_y = 0, \quad Y_x = 0, \quad Y_y = 0.$$

Furthermore, from (2.238) we obtain $U = 0$. Since Q is not quasilinear then from (2.233) we deduce that

$$(Xp + Yq)^2 + (Xq - Yp)^2 = 0,$$

from which we obtain $X = Y = 0$. With this assignment, we see from (2.231) that $F = Q$ and from (2.232) that Q satisfies

$$Q_{up} = 0, \qquad Q_{uq} = 0, \tag{2.239}$$

which has the solution

$$Q = G(p, q) + H(u) \tag{2.240}$$

for arbitrary functions G and H. From (2.229), we find that G satisfies $G_{pp} + G_{qq} = 0$; from (2.225d), that H satisfies $H'' = 0$ giving that $H = cu$ where c is an arbitrary constant, noting that we have suppressed the second constant of integration due to translational freedom. This leads to our main result. Equations of the form

$$u_t = u_{xx} + u_{yy} + cu + G\left(u_x, u_y\right),$$

are compatible with the first order equations

$$u_t = cu + G\left(u_x, u_y\right),$$

where c is an arbitrary constant and $G(p, q)$, a function satisfying $G_{pp} + G_{qq} = 0$.

2.4   Compatibility for Systems of PDEs

We now extend the idea of compatibility to systems of PDEs. In this section we consider the Cubic Schrodinger equation

$$i\psi_t + \psi_{xx} + k|\psi|^2\psi = 0. \tag{2.241}$$

If we assume that $\psi = u + iv$ then (2.241) becomes

$$\begin{aligned}
-v_t + u_{xx} + ku(u^2 + v^2) &= 0, \\
u_t + v_{xx} + kv(u^2 + v^2) &= 0.
\end{aligned} \tag{2.242}$$

Here, we seek compatibility with the pair of first order PDEs

$$\begin{aligned}
u_t + A(t, x, u, v)u_x + B(t, x, u, v)v_x &= U(t, x, u, v), \\
u_t + C(t, x, u, v)u_x + D(t, x, u, v)v_x &= V(t, x, u, v),
\end{aligned} \tag{2.243}$$

for some functions A, B, C, D, U and V to be determined. We solve (2.242) and (2.243) for u_t, v_t, u_{xx} and v_{xx} and require compatibility, that is

$$(u_t)_{xx} = (u_{xx})_t \quad \text{and} \quad (v_t)_{xx} = (v_{xx})_t. \tag{2.244}$$

Isolating the coefficients of u_x and v_x gives rise to two sets of determining equations for the unknowns A, B, C, D, U and V. Each set contains 10 determining equations. We only list 6 of each 10 as they are the only ones needed in our preliminary analysis. Also, they are the smaller equations. These are:

$$A_{uu} = 0, \quad (2.245\text{a})$$
$$2A_{uv} + B_{uu} = 0, \quad (2.245\text{b})$$
$$A_{vv} + 2B_{uv} = 0, \quad (2.245\text{c})$$
$$B_{vv} = 0, \quad (2.245\text{d})$$
$$U_{uu} - 2A_{xu} + (2C - B)A_u - AA_v - 2AB_u - DC_u - CC_v = 0, \quad (2.245\text{e})$$
$$U_{vv} - 2B_{xv} + 2DA_v + DB_u - (C + 4B)B_v - BD_u + (A - 2D)D_v = 0, \quad (2.245\text{f})$$
$$2U_{uv} - 2A_{xv} - 2B_{xu} + 3DA_u + (C - 2B)A_v - 3BB_u - 3AB_v$$
$$- BC_u + (A - 2D)C_v - DD_u - CD_v = 0, \quad (2.245\text{g})$$

and

$$C_{uu} = 0, \quad (2.246\text{a})$$
$$2C_{uv} + D_{uu} = 0, \quad (2.246\text{b})$$
$$C_{vv} + 2D_{uv} = 0, \quad (2.246\text{c})$$
$$D_{vv} = 0, \quad (2.246\text{d})$$
$$V_{uu} - 2C_{xu} + (2A - D)A_u + CA_v + (B + 4C)C_u - AC_v - 2AD_u = 0, \quad (2.246\text{e})$$
$$V_{vv} - 2D_{xv} + BB_u + AB_v + 2DC_v + DD_u + (C - 2B)D_v = 0, \quad (2.246\text{f})$$
$$2V_{uv} - 2C_{xv} - 2D_{xu} + BA_u + AA_v + (2A - D)B_u + CB_v$$
$$+ 3DC_u + 3CC_v + (2C - B)D_u - 3AD_v = 0. \quad (2.246\text{g})$$

We easily solve (2.245a)–(2.245d) and (2.246a)–(2.246d) giving

$$A = A_1(t, x)uv - B_1(t, x)v^2 + A_2(t, x)u + A_3(t, x)v + A_4(t, x), \quad (2.247\text{a})$$
$$B = -A_1(t, x)u^2 + B_1(t, x)uv + B_2(t, x)u + B_3(t, x)v + B_4(t, x), \quad (2.247\text{b})$$
$$C = C_1(t, x)uv - D_1(t, x)v^2 + C_2(t, x)u + C_3(t, x)v + C_4(t, x), \quad (2.247\text{c})$$
$$D = -C_1(t, x)u^2 + D_1(t, x)uv + D_2(t, x)u + D_3(t, x)v + D_4(t, x), \quad (2.247\text{d})$$

where A_i, B_i, C_i and $D_i, i = 1, 2, 3, 4$ are arbitrary functions of integration. We impose (2.247) on the remaining equations in (2.245) and (2.246) and then require the resulting equations to be compatible in the sense that

$$(U_{uu})_v = (U_{uv})_u, \quad (U_{uv})_v = (U_{vv})_u, \quad (V_{uu})_v = (V_{uv})_u, \quad \text{and} \quad (V_{uv})_v = (V_{vv})_u.$$
$$(2.248)$$

We performing this calculation and set the coefficients with respect to u, v, u^2, uv and v^2 to zero. The coefficients of u^2, uv and v^2 are presented below:

$$A_1^2 = 0, \tag{2.249a}$$

$$2A_1B_1 - A_1C_1 - 2C_1D_1 = 0, \tag{2.249b}$$

$$3A_1D_1 - B_1C_1 + B_1^2 + 4D_1^2 = 0, \tag{2.249c}$$

$$B_1D_1 = 0, \tag{2.250a}$$

$$6A_1B_1 - 3A_1C_1 + 4C_1D_1 = 0, \tag{2.250b}$$

$$3B_1^2 + 2D_1^2 - A_1D_1 - 3C_1B_1 = 0, \tag{2.250c}$$

$$A_1C_1 = 0, \tag{2.251a}$$

$$6C_1D_1 - 3B_1D_1 + 4A_1B_1 = 0, \tag{2.251b}$$

$$-2A_1^2 - 3C_1^2 + A_1D_1 + 3B_1C_1 = 0 = 0, \tag{2.251c}$$

and

$$D_1^2 = 0, \tag{2.252a}$$

$$2A_1B_1 + B_1D_1 - 2C_1D_1 = 0, \tag{2.252b}$$

$$4A_1^2 + C_1^2 + 3A_1D_1 - B_1C_1 = 0. \tag{2.252c}$$

From (2.249)–(2.252) we see that $A_1 = D_1 = 0$ and

$$B_1(B_1 - C_1) = 0, \quad C_1(B_1 - C_1) = 0, \tag{2.253}$$

from which we deduce that $C_1 = B_1$. We now return to (2.248) and isolate coefficients with respect to u and v. These are:

$$B_1(-7A_3 - 11B_2 - 3C_2 + 3D_3) = 0, \tag{2.254a}$$

$$B_1(-3A_2 + 2C_3 + D_2) = 0, \tag{2.254b}$$

$$B_1(5B_3 - C_3 + 2D_2) = 0, \tag{2.254c}$$

$$B_1(5A_3 + 9B_2 + C_2 - 9D_3) = 0, \tag{2.254d}$$

$$B_1(9A_2 - B_3 - 9C_3 - 5D_2) = 0, \tag{2.254e}$$

$$B_1(-2A_3 + B_2 - 5C_2) = 0, \tag{2.254f}$$

$$B_1(-A_3 - 2B_2 + 3D_3) = 0, \tag{2.254g}$$

$$B_1(-3A_2 + 3B_3 + 11C_3 + 7D_2) = 0, \tag{2.254h}$$

from which we see two cases emerge: (i) $B_1 \neq 0$ and (ii) $B_1 = 0$.

Case (i) $B_1 \neq 0$
Solving (2.254) gives

$$A_2 = 0, \ A_3 = -2B_2, \ C_1 = B_1, \ C_2 = B_2, \ C_3 = B_3, \ D_2 = -2B_3, \ D_3 = 0, \quad (2.255a)$$

and from the remaining compatibility conditions (2.248), we obtain

$$B_4 = \frac{B_2 B_3}{B_1} - \frac{3}{2} \frac{B_{1x}}{B_1}, \quad C_4 = \frac{B_2 B_3}{B_1} + \frac{3}{2} \frac{B_{1x}}{B_1}, \quad D_4 = A_4 + \frac{B_2^2 - B_3^2}{B_1}. \quad (2.256)$$

Now that we have consistency, we solve (2.245e)–(2.245g) and (2.246e)–(2.246g) for U and V. As these are fairly complicated (with 20 terms each), we suppress the output. We now return to (2.244) and consider the remaining 4 and 4 terms. As we are able to isolate terms involving u and v, we do so. This gives rise to a total of 114 new determining equations. It is left as an exercise to the reader to show that the final forms of A, B, C, D, U, and V are:

$$A = -c_1 v^2 + 2c_2 t + 2c_3, \quad B = C = c_1 uv, \quad D = -c_1 u^2 + 2c_2 t + 2c_3,$$
$$U = c_1 (c_2 t + c_3) (u^2 + v^2)v - (c_2 x + c_4)v, \quad (2.257)$$
$$V = -c_1 (c_2 t + c_3) (u^2 + v^2)u + (c_2 x + c_4)u,$$

giving rise to the following compatible equations:

$$u_t + (-c_1 v^2 + 2c_2 t + 2c_3) u_x + c_1 uvv_x = c_1(c_2 t + c_3)v(u^2 + v^2) - (c_2 x + c_4)v,$$
$$v_t + c_1 uvu_x + (-c_1 u^2 + 2c_2 t + 2c_3)v_x = -c_1(c_2 t + c_3)u(u^2 + v^2) + (c_2 x + c_4)u, \quad (2.258)$$

where $c_1 - c_4$ are arbitrary constants.

Case (ii) $B_1 = 0$
In order to facilitate finding a solution to (2.245) and (2.246) with the forms of A, B, C, and D given in (2.247) (with $A_1 = B_1 = C_1 = D_1 = 0$), we note that the original system (2.242) is invariant under the transformation

$$u \to -v, \quad v \to u. \quad (2.259)$$

We would expect the augmented pair (2.243) to also be invariant under the same transformation. In this case system (2.243) is

$$u_t + (A_2 u + A_3 v + A_4)u_x + (B_2 u + B_3 v + B_4)v_x = U(t, x, u, v),$$
$$v_t + (C_2 u + C_3 v + C_4)u_x + (D_2 u + D_3 v + D_4)v_x = V(t, x, u, v), \quad (2.260)$$

which becomes under (2.259)

$$u_t + (D_3 u - D_2 v + D_4)u_x + (-C_3 u + C_2 v - C_4)v_x = V(t, x, -v, u),$$
$$v_t + (-B_3 u + B_2 v - B_4)u_x + (A_3 u - A_2 v + A_4)v_x = -U(t, x, -v, u). \quad (2.261)$$

Comparing (2.260) and (2.261) gives

$$A_2 = D_3, \quad A_3 = -D_2, \quad A_4 = D_4, \quad B_2 = -C_3, \quad B_3 = C_2, \quad B_4 = -C_4,$$
$$C_2 = -B_3, \quad C_3 = B_2, \quad C_4 = -B_4, \quad D_2 = A_3, \quad D_3 = -A_2, \quad D_4 = A_4, \tag{2.262}$$

and from (2.262) we have

$$A_2 = B_2 = C_2 = D_2 = A_3 = B_3 = C_3 = D_3 = 0, \quad \text{and} \quad C_4 = -B_4, D_4 = A_4. \tag{2.263}$$

With these assignments (2.245e)–(2.245g) and (2.246e)–(2.246g) reduce considerably and the entire set of determining equations $(10 + 10)$ can be fully integrated, eventually leading to the compatible equations

$$u_t + \frac{c_1 x + 2c_2 t + c_3}{2c_1 t + c_0} u_x = \frac{-c_1 u - (c_2 x - c_4) v}{2c_1 t + c_0},$$
$$v_t + \frac{c_1 x + 2c_2 t + c_3}{2c_1 t + c_0} v_x = \frac{(c_2 x - c_4) u - c_1 v}{2c_1 t + c_0}, \tag{2.264}$$

where c_0–c_4 are arbitrary constants.

This chapter has considered the compatibility between PDEs; either single PDEs or systems of PDEs. It is interesting to note that several authors (see, for example, Pucci and Saccomandi [7], Arrigo and Beckham [8] and Niu et al. [9]) have shown that compatibility of a given PDE and a first order quasilinear PDE is equivalent to the nonclassical method in the symmetry analysis of differentials (Bluman and Cole [10]). Symmetry analysis of differential equations, first introduced by Lie [11], plays a fundamental role in the construction of exact solutions to nonlinear partial differential equations and provides a unified explanation for the seemingly diverse and ad-hoc integration methods used to solve ordinary differential equations. At the present time, there is extensive literature on the subject, and we refer the reader to the books by Arrigo [12], Bluman and Kumei [13], Cherniha et al. [14], and Olver [15] to name just a few.

2.5   Exercises

1. Solve the following PDEs by the method of characteristics:

$$(i) \quad u_x^2 - 3u_y^2 - u = 0, \quad u(x, 0) = x^2,$$
$$(ii) \quad u_t + u_x^2 + u = 0, \quad u(x, 0) = x,$$
$$(iii) \quad u_x^2 + u_y^2 = 1, \quad u(x, 1) = \sqrt{x^2 + 1},$$
$$(iv) \quad u_x^2 - u\, u_y = 0, \quad u(x, y) = 1 \text{ along } y = 1 - x.$$

2. Use Charpit's method to find compatible first order PDEs for the following

$$(i) \quad u_x^2 + u_y^2 = 2x,$$

$$(ii) \quad u_t + uu_x^2 = 0.$$

Use any one of the compatible equations derived above to obtain an exact solution of the PDEs given.

3. Show the PDE

$$u_t = (uu_x)_x$$

is compatible with

$$u_{xxx} = 0. \tag{2.267}$$

Further show that it admits solutions of the form

$$u = a(t)x^2 + b(t)x + c(t).$$

4. Find the value of k such that the PDE

$$u_t = uu_{xx} + ku_x^2$$

is compatible with

$$(i) \quad u_{xxxx} = 0,$$

$$(ii) \quad u_{xxxxx} = 0.$$

Further show that the original PDE admits solutions of the form

$$(i) \quad u = a(t)x^3 + b(t)x^2 + c(t)x + d(t),$$

$$(ii) \quad u = a(t)x^4 + b(t)x^3 + c(t)x^2 + d(t)x + e(t).$$

5. Show that

$$u_t = u_{xx} + u \ln u, \quad u_{tx} = \frac{u_t u_x}{u},$$

are compatible. Further show that the first PDE admits solutions of the form $u = T(t)X(x)$.

6. Show that

$$u_t = u_{xx} + u_x^2 + u^2, \quad u_{xxx} + u_x = 0,$$

are compatible. Further show that the first PDE admits solutions of the form $u = a(t) \cos x + b(t)$ [16].

7. Consider the compatibility between

$$u_t = D(u)u_{xx} \quad \text{and} \quad u_t = F(u)u_x^n. \tag{2.268}$$

Can you identify a relationship between $D(u)$ and $F(u)$ and for what powers n, such that these two PDEs are compatible?

8. In Sect. 3.4, we considered the Cubic Schrödiner equation

$$i\psi_t + \psi_{xx} + k|\psi|^2\psi = 0. \tag{2.269}$$

Assuming that

$$\psi = r \exp i\theta \tag{2.270}$$

then (2.269) becomes

$$r_t + r\theta_{xx} + 2r_x\theta_x = 0,$$
$$r\theta_t - r_{xx} + r\theta_x^2 - kr^3 = 0. \tag{2.271}$$

Perform compatibility between (2.271) and

$$r_t + E(t, x, r, \theta)r_x + F(t, x, r, \theta)\theta_x = R(t, x, r, \theta),$$
$$\theta_t + G(t, x, r, \theta)r_x + H(t, x, r, \theta)\theta_x = S(t, x, r, \theta), \tag{2.272}$$

and compare your results with those derived in Sect. 3.4. Note, the relations

$$u = r\cos\theta, \quad v = r\sin\theta, \tag{2.273}$$

will be needed to transform between the compatible equations (2.258) and (2.264) and (2.272).

References

1. D.J. Arrigo, *An Introduction to Partial Differential Equations* (Morgan Claypool, 2017)
2. D.J. Arrigo, Nonclassical contact symmetries and Charpit's method of compatibility. J. Non Math Phys. **12**(3), 321–329 (2005)
3. D.J. Arrigo, L.R. Suazo, First-order compatibility for a (2 + 1)-dimensional diffusion equation. J. Phys. A: Math. Theor. **41**, 025001 (2008)
4. A.C. Newell, J.A. Whitehead, Finite bandwidth, finite amplitude convection. J. Fluid Mech. **38**(2), 279–303 (1969)
5. L.A. Segel, Distant side-walls cause slow amplitude modulation of cellular convection. J. Fluid Mech. **38**, 203–224 (1969)
6. A.D. Polyanin, V.F. Zaitsev, Exact solutions of the Newell-Whitehead Equation from http://eqworld.ipmnet.ru
7. E. Pucci, G. Saccomandi, On the weak symmetry groups of partial differential equations. J. Math. Anal. Appl. **163**, 588–598 (1992)
8. D.J. Arrigo, J.R. Beckham, Nonclassical symmetries of evolutionary partial differential equations and compatibility. J. Math. Anal. Appl. **289**, 55–65 (2004)
9. X. Niu, L. Huang, Z. Pan, The determining equations for the nonclassical method of the nonlinear differential equation(s) with arbitrary order can be obtained through the compatibility. J. Math. Anal. Appl. **320**, 499–509 (2006)

10. G.W. Bluman, J.D. Cole, The general similarity solution of the heat equation. J. Math. Mech. **18**, 1025–1042 (1969)
11. S. Lie, Klassifikation und Integration von gewohnlichen Differentialgleichen zwischen x, y die eine Gruppe von Transformationen gestatten. Math. Ann. **32**, 213–281 (1888)
12. D.J. Arrigo, *Symmetries Analysis of Differential Equations–An Introduction* (Wiley, Hoboken, NJ, USA, 2015)
13. G. Bluman, S. Kumei, *Symmetries and Differential Equations* (Springer, Berlin, Germany, 1989)
14. R. Cherniha, S. Mykola, O. Pliukhin, *Nonlinear Reaction-Diffusion-Convection Equations: Lie and Conditional Symmetry, Exact Solutions and Their Applications* (CRC Press, Boca Raton, FL, USA, 2018)
15. P.J. Olver, *Applications of Lie Groups to Differential Equations*, 2nd edn. (Springer, Berlin, Germany, 1993)
16. V.A. Galaktionov, On new exact blow-up solutions for nonlinear heat conduction equations with source and applications. Diff. Int. Eqs. **3**, 863–874 (1990)

Differential Substitutions

3

One of the simplest nonlinear PDEs is Burgers' equation or the Bateman-Burgers' equation

$$u_t + u u_x = u_{xx}. \tag{3.1}$$

The PDE was introduced by Bateman in 1915 [2] and was used by Burgers' [1] as a rough model for turbulence. This equation has the steepening effect of the nonlinear PDE

$$u_t + u u_x = 0, \tag{3.2}$$

and the diffusive effect of the heat equation

$$u_t = u_{xx}. \tag{3.3}$$

Here, we first consider a related PDE—potential Burgers' equation. If we let

$$u = v_x, \tag{3.4}$$

then (3.1) can be integrated once, giving

$$v_t + \frac{1}{2} v_x^2 = v_{xx}. \tag{3.5}$$

Note that the function of integration can be set to zero with out loss of generality. A remarkable fact is that this nonlinear PDE (3.5) can be linearized. If we let

$$v = f(w), \tag{3.6}$$

then (3.5) becomes

$$f'(w) w_t + \frac{1}{2} f'^2(w) w_x^2 = f'(w) w_{xx} + f''(w) w_x^2. \tag{3.7}$$

D. Arrigo, *Analytical Methods for Solving Nonlinear Partial Differential Equations*,
Synthesis Lectures on Mathematics & Statistics.
https://doi.org/10.1007/978-3-031-17069-0_3

Choosing

$$f'' = \frac{1}{2} f'^2 \tag{3.8}$$

gives (3.7) as the heat equation

$$w_t = w_{xx}! \tag{3.9}$$

Solving (3.8) gives

$$f = -2 \ln |w + c_1| + c_2, \tag{3.10}$$

where c_1 and c_2 are constants of integration. Setting $c_1 = c_2 = 0$ then $f = -2 \ln |w|$ and from (3.6)

$$v = -2 \ln |w|, \tag{3.11}$$

transforming the potential Burgers' equation (3.5)–(3.9) the heat equation. Now we return to the actual Burgers equation (3.1). Combining (3.4) and (3.11) tells us that Burgers' equation and the heat equation are related via

$$u = -2 \frac{w_x}{w}. \tag{3.12}$$

This transformation is known as the Hopf-Cole transformation, introduced independently by Hopf [3] and Cole [4], but appeared earlier in Forsythe [5]. If we start with Burgers' equation (3.1), we can see that it is transformed to the heat equation through the Hope-Cole transformation (3.12). The following illustrates:

$$u_t = \left(u_x - \frac{1}{2} u^2 \right)_x$$

$$\left(-2 \frac{w_x}{w} \right)_t = \left(-2 \frac{w_{xx}}{w} + 2 \frac{w_x^2}{w^2} - \frac{1}{2} \left(-2 \frac{w_x}{w} \right)^2 \right)_x$$

$$\left(\frac{w_x}{w} \right)_t = \left(\frac{w_{xx}}{w} \right)_x \tag{3.13}$$

$$\left(\frac{w_t}{w} \right)_x = \left(\frac{w_{xx}}{w} \right)$$

$$w_t = w_{xx} + c'(t) w$$

where $c'(t)$ is arbitrary. However, we can set $c'(t) = 0$ as introducing the new variable $\tilde{w} = e^{c(t)} w$, gives the last of (3.13) with $c'(t) = 0$ and further leaves (3.12) unchanged. We now consider two examples.

Example 3.1 Solve the initial value problem

$$u_t + u u_x = u_{xx}, \quad 0 < x < \pi,$$
$$u(0, t) = u(\pi, t) = 0, \quad u(x, 0) = \sin x. \tag{3.14}$$

Passing the PDE, the boundary conditions, and the initial condition through the Hopf-Cole transformation (3.12) gives

$$w_t = w_{xx}$$
$$w_x(0, t) = w_x(\pi, t) = 0 \tag{3.15}$$
$$w(t, 0) = \exp\left(\frac{1}{2}\cos x\right).$$

With the usual separation of variables, we find the solution of (3.15)

$$w = \frac{1}{2}a_0 + \sum_{n=1}^{\infty} a_n e^{-n^2 t} \cos nx \tag{3.16}$$

where

$$a_n = \frac{2}{\pi}\int_0^{\pi} \exp\left(\frac{1}{2}\cos x\right)\cos nx\, dx = 2I_n\left(\frac{1}{2}\right) \tag{3.17}$$

and $I_n(x)$ is the modified Bessel function of the first kind. Passing (3.16) through the Hopf-Cole transformation (3.12) gives

$$u = \frac{4\sum_{n=1}^{\infty} n I_n\left(\frac{1}{2}\right)e^{-n^2 t}\sin nx}{I_0(\frac{1}{2}) + 2\sum_{n=1}^{\infty} I_n(\frac{1}{2})e^{-n^2 t}\cos nx}. \tag{3.18}$$

If we solve the heat equation with the same boundary and initial conditions as given in (3.14), we obtain the simple solution

$$u = e^{-t}\sin x. \tag{3.19}$$

Figure 3.1 shows a comparison of the solutions (3.18) and (3.19) at a variety of times.

Example 3.2 Solve the initial value problem

$$u_t + u u_x = u_{xx}, \quad \infty < x < \infty, \quad u(t, 0) = f(x) \tag{3.20}$$

Passing the initial condition through the Hopf-Cole transformation gives

$$w(t, 0) = W(x) = \exp\left(-\frac{1}{2}\int_0^x f(\xi)\, d\xi\right). \tag{3.21}$$

The solution of the heat equation (3.9) with this initial condition is

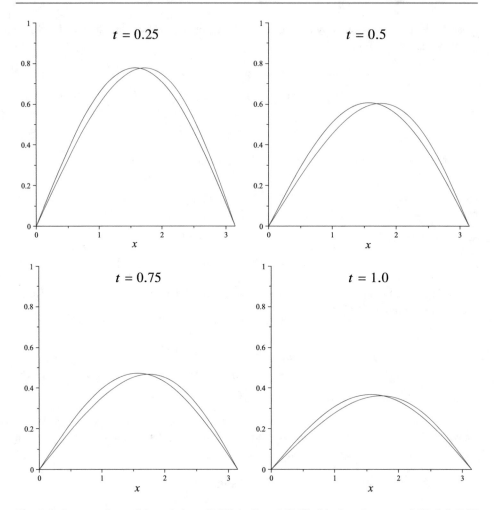

Fig. 3.1 A comparison of the solutions (3.18) (red) and (3.19) (blue) at times t = 0.25, 0.5, 0.75 and 1

$$w(x, t) = \frac{1}{\sqrt{4\pi t}} \int_{-\infty}^{\infty} W(\xi)\, e^{-(x-\xi)^2/4t}\, d\xi. \tag{3.22}$$

from which we obtain

$$w_x(x, t) = -\frac{1}{\sqrt{4\pi t}} \int_{-\infty}^{\infty} \frac{(x-\xi)}{2t} W(\xi) e^{-(x-\xi)^2/4t}\, d\xi. \tag{3.23}$$

From (3.12) we find the following

$$u(x,t) = \frac{\displaystyle\int_{-\infty}^{\infty} \frac{(x-\xi)}{2t} W(\xi) e^{-(x-\xi)^2/4t}\, d\xi}{\displaystyle\int_{-\infty}^{\infty} W(\xi) e^{-(x-\xi)^2/4t}\, d\xi}. \tag{3.24}$$

As a particular example, we consider

$$f(x) = \begin{cases} 0 & x < -1 \\ 2k & -1 \le x \le 1 \\ 0 & x > 1 \end{cases} \tag{3.25}$$

where k is a positive constant. The initial condition (3.21) becomes

$$w(x,0) = \begin{cases} e^k & x < -1 \\ e^{-kx} & -1 \le x \le 1 \\ e^{-k} & x > 1. \end{cases} \tag{3.26}$$

With this initial condition, (3.22) becomes

$$\begin{aligned}
w(x,t) &= \frac{e^k}{2}\left(1 - \operatorname{erf}\left(\frac{x+1}{2\sqrt{t}}\right)\right) + \frac{e^{-k}}{2}\left(1 + \operatorname{erf}\left(\frac{x-1}{2\sqrt{t}}\right)\right) \\
&\quad - \frac{e^{k^2 t - kx}}{2}\left(\operatorname{erf}\left(\frac{2kt - x - 1}{2\sqrt{t}}\right) - \operatorname{erf}\left(\frac{2kt - x + 1}{2\sqrt{t}}\right)\right),
\end{aligned} \tag{3.27}$$

and through the Hopf-Cole transformation (3.12) we obtain the solution to (3.20), subject to the initial condition (3.25) .

3.1 Generalized Burgers' Equation

Here we consider a generalization of the Hopf-Cole transformation. Can we connect the following PDEs

$$u_t + A(u)u_x = u_{xx}, \tag{3.28a}$$

$$v_t + B(v)v_x = v_{xx}, \tag{3.28b}$$

via the substitution

$$u = F(v, v_x), \quad F_{v_x} \ne 0? \tag{3.29}$$

Substituting (3.29) into (3.28a) and imposing (3.28b) leads to the following determining equations:

$$F_{pp} = 0, \tag{3.30a}$$

$$A(F)F_p - B(v)F_p - 2pF_{vp} = 0, \tag{3.30b}$$

$$A(F)F_v - pB(v)F_p - p^2 B' F_p - p^2 F_{vv} = 0, \tag{3.30c}$$

where $p = v_x$. Differentiating (3.30b) with respect to p twice (using (3.30a)) gives

$$F_p^3 A''(F) = 0. \tag{3.31}$$

From (3.31) and (3.30a) we find

$$A(F) = c_1 F + c_2, \quad F = F_1(v)p + F_2(v), \tag{3.32}$$

where c_1 and c_2 are arbitrary constants and F_1 and F_2 are arbitrary functions. With these assignments, returning to (3.30b) and isolating coefficients with respect to p gives

$$2F_1' - c_1 F_1^2 = 0, \tag{3.33a}$$

$$F_1 (c_1 F_2 + c_2 - B) = 0, \tag{3.33b}$$

and since $F_1 \neq 0$, then we solve giving

$$F_1 = \frac{-2}{c_1 v + c_3}, \quad B = c_1 F_2 + c_2, \tag{3.34}$$

where c_3 is an additional arbitrary constant. The remaining equation in (3.30c) becomes

$$F_2'' = 0, \tag{3.35}$$

which we solve as

$$F_2 = c_4 v + c_5. \tag{3.36}$$

Composing our results and invoking a simple translation and scaling of the variables u and v we have that solutions of

$$u_t + u u_x = u_{xx} \tag{3.37}$$

can be obtain from

$$v_t + (av + b)v_x = v_{xx} \tag{3.38}$$

via the substitution

$$u = -2\frac{v_x}{v} + av + b, \tag{3.39}$$

where a and b are arbitrary constant. If we set $a = b = 0$ we get the Hopf-Cole transformation. It is interesting to note that if we set $a = 1, b = 0$, we get a transformation which gives rise to solutions of the same equation.

3.2   KdV-MKdV Connection

The Korteweg-deVries equation (KdV)

$$u_t + 6uu_x + u_{xxx} = 0, \tag{3.40}$$

first introduced by Korteweg and DeVries [6] to model shallow water waves, is a remarkable PDE. It has a number of applications and possesses a number of special properties (see, for example, Miura [7]). In 1968 Robert Miura found this remarkable transformation [8]. He found that solutions of the KdV equation (3.40) can be found using solutions of the modified Korteweg-deVries equation

$$v_t - 6v^2 v_x + v_{xxx} = 0, \tag{3.41}$$

via the transformation

$$u = v_x - v^2, \tag{3.42}$$

which today is known as the Miura transformation. We ask whether it's possible to connect two general KdV type equations.

3.3   Generalized KdV Equation

In this section we connect the following general PDEs

$$u_t + A(u)u_x + u_{xxx} = 0, \tag{3.43a}$$
$$v_t + B(v)v_x + v_{xxx} = 0, \tag{3.43b}$$

via the substitution

$$u = F(v, v_x), \quad F_{v_x} \neq 0. \tag{3.44}$$

Substituting (3.44) into (3.43a) and imposing (3.43b) leads to the following determining equations:

$$F_{pp} = 0, \tag{3.45a}$$
$$F_{vp} = 0, \tag{3.45b}$$
$$A(F)F_p - B(v)F_p + 3pF_{vv} = 0, \tag{3.45c}$$
$$A(F)F_v - B(v)F_p - pB'F_p + p^2 F_{vvv} = 0, \tag{3.45d}$$

where $p = v_x$. Differentiating (3.45c) with respect to p twice (using (3.45a) and (3.45b)) gives

$$F_p^3 A''(F) = 0. \tag{3.46}$$

From (3.46), (3.45a) and (3.45b) we see

$$A(F) = c_1 F + c_2, \quad F = c_3 p + F_1(v), \tag{3.47}$$

where $c_0 - c_3$ are arbitrary constants and F_1 an arbitrary function. With these assignments, returning to (3.45c) and isolating coefficients with respect to p gives

$$3F_1'' + c_1 c_3^2 = 0, \tag{3.48a}$$
$$B - c_1 F_1 - c_2 = 0, \tag{3.48b}$$

which we solve as

$$F_1 = -\frac{c_1 c_3^2}{6} v^2 + c_4 v + c_5, \quad B = c_1 F_1 + c_2, \tag{3.49}$$

(c_4 and c_5 additional constants) and shows that (3.45d) is automatically satisfied. As we have the flexibility of translation and scaling in the variables, without loss of generality we can set $c_0 = 0, c_1 = 6, c_2 = 1, c_3 = a, c_4 = b$. Thus, solutions of

$$u_t + 6uu_x + u_{xxx} = 0 \tag{3.50}$$

can be obtained via

$$u = v_x - v^2 + av + b, \tag{3.51}$$

where v satisfies

$$v_t + 6(-v^2 + av + b)v_x + v_{xxx} = 0. \tag{3.52}$$

Setting $a = b = 0$ gives the Muira transformation (3.42); the usual KdV-MKdV connection, setting $b = 0$, gives a connection between the KdV and what is known as the Gardner equation [9].

Example 3.3 We consider the nonlinear PDE

$$u_{xx} + 2uu_{xy} + u^2 u_{yy} = 0. \tag{3.53}$$

This particular PDE arises in the study of highly frictional granular materials ([10]). We consider a differential substitution

$$u = F(v_x, v_y), \tag{3.54}$$

and the target PDE

$$v_{xx} + A(v, v_x, v_y)v_{xy} + B(v, v_x, v_y)v_{yy} + C(v, v_x, v_y) = 0. \tag{3.55}$$

Substitution of (3.54) into (3.53) and imposing (3.55), on isolating the coefficients of $v_{xy}, v_{yy}, v_{xyy}, v_{yyy}$ and various products gives rise to the following determining equations:

$$C^2 F_{pp} + (CC_p - pC_v + qAC_v - 2qFC_v)F_p - qC_v F_q = 0,$$
(3.56a)

$$2C(A - F)F_{pp} - 2CF_{pq} - (C_q + qA_v)F_q +$$
$$(qAA_v - pA_v - 2FC_p - 2qFA_v - C_q + CA_p + 2AC_p)F_p = 0,$$
(3.56b)

$$2BCF_{pp} - 2FCF_{pq} - (C_q + qB_v)F_q +$$
$$(AC_q - 2FC_q - pB_v + CB_p - 2qFB_v + BC_p + qAB_v)F_p = 0,$$
(3.56c)

$$(F - A)^2 F_{pp} + 2(F - A)F_{pq} + F_{qq} + (2AA_p - 2FA_p - A_q)F_p - A_p F_q = 0,$$
(3.56d)

$$2B(A - F)F_{pp} + 2(F^2 - AF - B)F_{pq} + 2FF_{qq} - (A_q + B_p)F_q +$$
$$(BA_p - B_q - 2FB_p - 2FA_q + AA_q + 2AB_p)F_p = 0,$$
(3.56e)

$$B^2 F_{pp} - 2FBF_{pq} + F^2 F_{qq} + (AB_q - 2FB_q + BB_p)F_p - B_q F_q = 0,$$
(3.56f)

$$(F^2 + A^2 - B - 2AF)F_p + (2F - A)F_q = 0,$$
(3.56g)

$$B(A - 2F)F_p + (F^2 - B)F_q = 0.$$
(3.56h)

Eliminating F_p from (3.56g) and (3.56h) gives

$$B = AF - F^2,$$
(3.57)

and returning to either (3.56g) and (3.56h) gives

$$(A - 2F)\left((F - A)F_p + F_q\right) = 0,$$
(3.58)

leading to two cases: (i) $A = 2F$ and (ii) $A = \dfrac{FF_p + F_q}{F_p}$, but each case ultimately leads to the same result and thus, we will only pursue the first case. On setting $A = 2F$ in Eqs. (3.56d), (3.56e) and (3.56f), we obtain

$$F^2 F_{pp} - 2FF_{pq} + F_{qq} + 4FF_p^2 - 4F_p F_q = 0, \tag{3.59a}$$

$$F^3 F_{pp} - 2F^2 F_{pq} + FF_{qq} + 3F^2 F_p^2 - 2FF_p F_q - F_q^2 = 0, \tag{3.59b}$$

$$F^4 F_{pp} - 2F^3 F_{pa} + F^2 F_{qq} + 2F^3 F_p^2 - 2FF_q^2 = 0, \tag{3.59c}$$

and on the elimination of the second order derivatives we obtain

$$F_q - FF_p = 0, \tag{3.60}$$

and with this shows that (3.59) is identically satisfied. As F satisfies (3.60), both (3.56b) and (3.56c) reduce to

$$C_q - FC_p = 0. \tag{3.61}$$

With F and C satisfying (3.59) and (3.61), and the last equation in (3.56), (3.56a) becomes

$$\left(CC_p - pC_v - qFC_v\right) F_p + C^2 F_{pp} = 0. \tag{3.62}$$

If we perform the operation $\partial_q - F\partial_p$ on (3.62) and use (3.60) and (3.61), we obtain

$$2CF_p C_p - (p+qF)F_p C_v + 3C^2 F_{pp} = 0. \tag{3.63}$$

From (3.62) and (3.63), (upon subtraction), we have

$$C\left(F_p C_p + 2CF_{pp}\right) = 0. \tag{3.64}$$

If $C = 0$ then we have

$$u = F(p,q), \quad v_{xx} + 2Fv_{xy} + F^2 v_{yy} = 0, \quad F_q - FF_p = 0. \tag{3.65}$$

If $C \neq 0$ then the compatibility of $F_p C_p + 2CF_{pp} = 0$ with (3.60) and (3.61) leads to

$$F_{pp} = 0, \tag{3.66}$$

and from (3.61) and (3.64) gives $C_p = C_q = 0$, and from (3.62) gives $(p+qF)C_v = 0$, which leads to two cases. If $F = -p/q$, then

$$u = -\frac{v_x}{v_y}, \quad v_y^2 v_{xx} - 2v_x v_y v_{xy} + v_x^2 v_{yy} + c(v)v_y^2 = 0, \tag{3.67}$$

where $c(v)$ is an arbitrary function. If $C_v = 0$, then

$$u = -\frac{v_x + c_1}{v_y + c_2}, \tag{3.68}$$

$$(v_y + c_2)^2 v_{xx} - 2(v_x + c_1)(v_y + c_2)v_{xy} + (v_x + c_1)^2 v_{yy} + c_3(v_y + c_2)^2 = 0.$$

where c_1 and c_2 are arbitrary constants. However, we can set $c_1 = c_2 = 0$ in (3.68) without loss of generality showing (3.68) to be a special case of (3.67).

A natural question one might ask is "what was the point?" We have shown that we are able to express solutions of one nonlinear PDE, (3.53), in terms of another nonlinear PDE, (3.65) or (3.67)—a rather esoteric exercise. However, as it turns out, these PDEs, ((3.65) and (3.67)) are linearizable! This is a topic covered in the next chapter.

3.4 Matrix Hopf-Cole Transformation

In this section we extend the Hopf-Cole transformation to matrices. Before doing so, we must establish the derivative of inverse matrices. Consider an $n \times n$ invertible matrix Φ with variable entries. Then

$$\Phi \, \Phi^{-1} = \mathbb{I} \tag{3.69}$$

where $\mathbb{I}$ is the usual identity matrix. Differentiating with respect to x gives

$$\Phi_x \, \Phi^{-1} + \Phi \, \Phi_x^{-1} = 0, \tag{3.70}$$

from which we obtain

$$\Phi_x^{-1} = -\Phi^{-1} \, \Phi_x \, \Phi^{-1}, \tag{3.71}$$

noting that $\Phi_x^{-1} = \left(\Phi^{-1}\right)_x$ and not $\Phi_x^{-1} = (\Phi_x)^{-1}$. Similarly for the t derivative

$$\Phi_t^{-1} = -\Phi^{-1} \, \Phi_t \, \Phi^{-1}. \tag{3.72}$$

We know consider the Matrix Burgers' equation

$$\Omega_t + 2\Omega_x \Omega = \Omega_{xx}, \tag{3.73}$$

and matrix Hopf-Cole transformation

$$\Omega = -\Phi_x \Phi^{-1}. \tag{3.74}$$

Substituting (3.74) into (3.73) (using (3.71) and (3.72) appropriately) gives

$$-\Phi_{tx}\Phi^{-1} + \Phi_x\Phi^{-1}\Phi_t\Phi^{-1} - 2\left(-\Phi_{xx}\Phi^{-1} + \Phi_x\Phi^{-1}\Phi_x\Phi^{-1}\right)\Phi_x\Phi^{-1} =$$
$$-\Phi_{xxx}\Phi^{-1} + 2\Phi_{xx}\Phi^{-1}\Phi_x\Phi^{-1} + \Phi_x\Phi^{-1}\Phi_{xx}\Phi^{-1} - 2\Phi_x\Phi^{-1}\Phi_x\Phi^{-1}\Phi_x\Phi^{-1}. \tag{3.75}$$

Expanding and cancelling appropriately gives

$$-\Phi_{tx}\Phi^{-1} + \Phi_x\Phi^{-1}\Phi_t\Phi^{-1} = -\Phi_{xxx}\Phi^{-1} + \Phi_x\Phi^{-1}\Phi_{xx}\Phi^{-1}, \tag{3.76}$$

which becomes on multiplying on the left by $-\Phi^{-1}$ and on the right by Φ

$$\left(\Phi^{-1}\Phi_t\right)_x = \left(\Phi^{-1}\Phi_{xx}\right)_x. \tag{3.77}$$

Integrating with respect to x gives the linear heat matrix equation

$$\Phi_t = \Phi_{xx}. \tag{3.78}$$

Note that, like the scalar case, the matrix function of integration can be set to zero without loss of generality.

3.5   Darboux Transformations

In 1882 Darboux considered the following problem: Is it possible to find functions $f(x)$ such that solutions of

$$y'' + f(x)y = 0, \quad z'' + kz = 0, \tag{3.79}$$

can be connected through

$$y = z' + A(x)z? \tag{3.80}$$

A direct substitution of (3.80) into the first of (3.79) gives

$$z''' + Az'' + 2A'z' + A''z + f(x)\left(z' + Az\right) = 0, \tag{3.81}$$

where prime denoted differentiation with respect to their argument. Imposing the second of (3.79) gives

$$\left(f + 2A' - k\right)z' + \left(A'' + fA - kA\right)z = 0, \tag{3.82}$$

and since this must be true for all z leads to

$$f + 2A' - k = 0, \tag{3.83a}$$
$$A'' + fA - kA = 0. \tag{3.83b}$$

If $f = k - 2A'$ then (3.83b) becomes

$$A'' - 2AA' = 0, \tag{3.84}$$

which integrates once to become

$$A' - A^2 = c \tag{3.85}$$

where c is an arbitrary constant. If we let $A = -\phi'/\phi$ where $\phi = \phi(x)$, (3.85) becomes

$$\phi'' + c\phi = 0. \tag{3.86}$$

Composing the result gives the solution of

$$y'' + \left(2\left(\ln \phi\right)'' + k\right)y = 0 \tag{3.87}$$

are obtained via the Darboux transformation

$$y = z' - \frac{\phi'}{\phi} z, \tag{3.88}$$

where ϕ and z satisfy

$$\phi'' + c\phi = 0, \quad z'' + kz = 0, \tag{3.89}$$

respectively.

For example, if we choose solutions $\phi = \cosh x$ and $z = ax + b$ (a, b constant) as solutions of (3.89), then via (3.87) and (3.88) solutions of

$$y'' + \frac{2y}{\cosh^2 x} = 0 \tag{3.90}$$

are given by

$$y = a - (ax + b) \tanh x. \tag{3.91}$$

A natural question is: Does this method apply to PDEs? The answer—Yes! A very important class of PDEs is the $(1 + 1)$ dimensional Schrödinger equation

$$i\psi_t = -\frac{\hbar}{2m} \psi_{xx} + V(x)\psi \tag{3.92}$$

from Quantum Mechanics. Here, we will consider

$$u_t = u_{xx} + f(t, x)u, \tag{3.93}$$

and seek Darboux transformations of the form

$$u = v_x + A(t, x)v, \tag{3.94}$$

where v satisfies the standard heat equation

$$v_t = v_{xx}. \tag{3.95}$$

Substituting (3.94) into (3.93) and imposing (3.95) gives

$$(A_t - A_{xx} - fA)\, v_x - (2A_x + f)\, v = 0. \tag{3.96}$$

Since (3.96) must be true for all v, we obtain

$$2A_x + f = 0, \tag{3.97a}$$
$$A_t - A_{xx} - fA = 0. \tag{3.97b}$$

From (3.97a), we find $f = -2A_x$, and eliminating f in (3.97b) gives

$$A_t + 2AA_x - A_{xx} = 0. \tag{3.98}$$

As we saw previously, this is Burgers' equation which was linearized via the Hopf-Cole transformation

$$A = -\phi_x/\phi. \tag{3.99}$$

This, in turn, gives $f = 2\,(\ln \phi)_{xx}$. Thus, we have the following: solutions of

$$u_t = u_{xx} + 2\,(\ln \phi)_{xx}\, u \tag{3.100}$$

are given by

$$u = v_x - \frac{\phi_x}{\phi} v \tag{3.101}$$

where both ϕ and v satisfy the heat equation. We now consider a few examples, choosing various simple solutions of the heat equation.

Example 3.4 If we choose $\phi = x$, then solutions of

$$u_t = u_{xx} - \frac{2u}{x^2} \tag{3.102}$$

are obtained by

$$u = v_x - \frac{v}{x}, \tag{3.103}$$

where here, and in the examples that follow, v is any solution of the heat equation.

Example 3.5 If we choose $\phi = x^2 + 2t$, then solutions of

$$u_t = u_{xx} - \frac{(x^2 - 2t)}{(x^2 + 2t)^2} u \tag{3.104}$$

are obtained by

$$u = v_x - \frac{2x}{x^2 + 2t} v. \tag{3.105}$$

Example 3.6 If we choose $\phi = e^t \cosh x$, then solutions of

$$u_t = u_{xx} + \frac{u}{\cosh^2 x} \tag{3.106}$$

are obtained by

$$u = v_x - \tanh x v. \tag{3.107}$$

An interesting question arises: Is it possible to use Darboux transformations for any time independent potential? See the exercise section for further examination.

3.5.1   Second Order Darboux Transformations

A natural extension is to consider a transformation that includes second derivatives, *i.e.*

$$u = v_{xx} + A(t, x)v_x + B(t, x)v. \tag{3.108}$$

Substituting (3.108) into into (3.93) and imposing (3.95) gives

$$-(2A_x + f)\, v_{xx} + (A_t - 2B_x - A_{xx} - f A)\, v_x + (B_t - B_{xx} - f B)\, v = 0. \tag{3.109}$$

As this must be satisfied for all v we obtain

$$2A_x + f = 0, \tag{3.110a}$$
$$A_t - 2B_x - A_{xx} - f A = 0, \tag{3.110b}$$
$$B_t - B_{xx} - f B = 0. \tag{3.110c}$$

We see again $f = -2A_x$, and the two remaining equations in (3.110) are

$$A_t + 2A A_x - 2B_x - A_{xx} = 0, \tag{3.111a}$$
$$B_t + 2B A_x - B_{xx} = 0. \tag{3.111b}$$

It is interesting to note that in order to find a second order Darboux transformation to solve a linear PDE, we must solve a system of nonlinear systems of PDEs—a task that might seem hopeless. However, we see that by introducing the matrix

$$\Omega = \begin{bmatrix} 0 & -1 \\ B & A \end{bmatrix}, \tag{3.112}$$

Burgers' system (3.111) can be written as

$$\Omega_t + 2\Omega_x \Omega - \Omega_{xx} = 0 \tag{3.113}$$

and, as we saw in the section on the matrix Burgers' equation, can be linearized via a Matrix Hopf-Cole transformation. Thus, if we let

$$\Omega = -\Phi_x \Phi^{-1} \tag{3.114}$$

where

$$\Phi = \begin{bmatrix} \phi_{11} & \phi_{12} \\ \phi_{21} & \phi_{22} \end{bmatrix}, \quad \phi_{ij} = \phi_{ij}(t, x), \quad i, j = 1, 2, \tag{3.115}$$

then (3.113) becomes

$$\Phi_t - \Phi_{xx} = 0. \tag{3.116}$$

Thus, each element of Φ satisfies the heat equation. From (3.112), (3.114) and (3.115), we have

$$- \Phi_x = \Omega\Phi, \tag{3.117}$$

which becomes

$$- \begin{bmatrix} \phi_{11_x} & \phi_{12_x} \\ \phi_{21_x} & \phi_{22_x} \end{bmatrix} = \begin{bmatrix} 0 & -1 \\ B & A \end{bmatrix} \begin{bmatrix} \phi_{11} & \phi_{12} \\ \phi_{21} & \phi_{22} \end{bmatrix}. \tag{3.118}$$

Expanding (3.118) and equating elements gives the following

$$\phi_{11_x} = \phi_{21}, \quad -\phi_{21_x} = B\phi_{11} + A\phi_{21},$$
$$\phi_{12_x} = \phi_{22}, \quad -\phi_{22_x} = B\phi_{12} + A\phi_{22}. \tag{3.119}$$

If we let $\phi_{11} = \omega_1$ and $\phi_{12} = \omega_2$, then $\phi_{21} = \omega_{1_x}$ and $\phi_{22} = \omega_2$ then from (3.119)

$$B\omega_1 + A\omega_{1_x} = -\omega_{1_{xx}}, \tag{3.120a}$$
$$B\omega_2 + A\omega_{2_x} = -\omega_{2_{xx}}, \tag{3.120b}$$

a system of two equations in the two unknowns A and B. Solving gives

$$A = - \frac{\begin{vmatrix} \omega_1 & \omega_2 \\ \omega_{1_{xx}} & \omega_{2_{xx}} \end{vmatrix}}{\begin{vmatrix} \omega_1 & \omega_2 \\ \omega_{1_x} & \omega_{2_x} \end{vmatrix}}, \quad B = \frac{\begin{vmatrix} \omega_{1_x} & \omega_{2_x} \\ \omega_{1_{xx}} & \omega_{2_{xx}} \end{vmatrix}}{\begin{vmatrix} \omega_1 & \omega_2 \\ \omega_{1_x} & \omega_{2_x} \end{vmatrix}}. \tag{3.121}$$

Furthermore, $f = -2A_x$ which can be written as

$$f = 2 \left(\ln W_2\right)_{xx} \quad \text{where} \quad W_2 = \omega_1\omega_{2_x} - \omega_2\omega_{1_x}. \tag{3.122}$$

Substituting A and B from (3.121) into (3.108) gives

$$u = v_{xx} - \frac{\begin{vmatrix} \omega_1 & \omega_2 \\ \omega_{1_{xx}} & \omega_{2_{xx}} \end{vmatrix}}{\begin{vmatrix} \omega_1 & \omega_2 \\ \omega_{1_x} & \omega_{2_x} \end{vmatrix}} v_x + \frac{\begin{vmatrix} \omega_{1_x} & \omega_{2_x} \\ \omega_{1_{xx}} & \omega_{2_{xx}} \end{vmatrix}}{\begin{vmatrix} \omega_1 & \omega_2 \\ \omega_{1_x} & \omega_{2_x} \end{vmatrix}} v \tag{3.123}$$

which can conveniently be written as

$$u = \frac{\begin{vmatrix} \omega_1 & \omega_2 & v \\ \omega_{1_x} & \omega_{2_x} & v_x \\ \omega_{1_{xx}} & \omega_{2_{xx}} & v_{xx} \end{vmatrix}}{\begin{vmatrix} \omega_1 & \omega_2 \\ \omega_{1_x} & \omega_{2_x} \end{vmatrix}}, \tag{3.124}$$

or

$$u = \frac{|W(\omega_1, \omega_2, v)|}{|W(\omega_1, \omega_2)|}, \tag{3.125}$$

where W is the usual Wronskian.

Example 3.7 If we choose $\omega_1 = x$, and $\omega_2 = x^3 + 6xt$, then from (3.121) and (3.122)

$$f = -\frac{6}{x^2}, \quad A = -\frac{3}{x}, \quad B = \frac{3}{x^2}, \tag{3.126}$$

and solutions of

$$u_t = u_{xx} - \frac{6u}{x^2} \tag{3.127}$$

are obtained by

$$u = v_{xx} - \frac{3v_x}{x} + \frac{3v}{x^2}. \tag{3.128}$$

Example 3.8 If we choose $\omega_1 = e^{a^2 t} \cosh ax$, and $\omega_2 = e^{b^2 t} \sinh bx$, where a and b are constants, then from (3.121) and (3.122)

$$
\begin{aligned}
f &= -2\frac{(a^2 - b^2)(b^2 \cosh^2 ax + a^2 \cosh^2 bx)}{(a \sinh ax \sinh bx - b \cosh ax \cosh bx)^2}, \\
A &= -\frac{(a^2 - b^2) \cosh ax \sinh bx}{a \sinh ax \sinh bx - b \cosh ax \cosh bx}, \\
B &= -\frac{ab(a \cosh ax \cosh bx - b \sinh ax \sinh bx)}{a \sinh ax \sinh bx - b \cosh ax \cosh bx},
\end{aligned} \tag{3.129}
$$

and solutions of

$$u_t = u_{xx} - 2\frac{(a^2 - b^2)(b^2 \cosh^2 ax + a^2 \cosh^2 bx)}{(a \sinh ax \sinh bx - b \cosh ax \cosh bx)^2} u \tag{3.130}$$

can be obtained using (3.108) with A and B given in (3.129). For example, choosing $a = 1$ and $b = 2$ shows that PDEs of the form

$$u_t = u_{xx} + \frac{6u}{\cosh^2 x}, \tag{3.131}$$

admit a Darboux transformation of the form

$$u = v_{xx} - \frac{3 \sinh x}{\cosh x} v_x + \frac{2 \cosh^2 x - 3}{\cosh^2 x} v. \tag{3.132}$$

These results generalize to nth order Darboux transformations; we refer the reader to Arrigo and Hickling [11] for futher details.

3.5.2 Darboux Transformations Between Two Diffusion Equations

In this section we wish to extend the previous section's results and now have v satisfy

$$v_t = v_{xx} + g(t, x)v. \tag{3.133}$$

3.5.2.1 First Order Darboux Transformations

We will start with first order Darboux transformations as we did earlier and seek solutions of

$$u_t = u_{xx} + f(t, x)u \tag{3.134}$$

via the Darboux transformations of the form

$$u = v_x + A(t, x)v, \tag{3.135}$$

where v satisfies (3.133). As we did previously, substituting (3.135) into (3.134), imposing (3.133) and isolating coefficients with respect to v and v_x, gives

$$2A_x + f - g = 0, \tag{3.136a}$$
$$A_t - A_{xx} - fA + g_x + gA = 0. \tag{3.136b}$$

From (3.136a) we see that
$$f = g - 2A_x, \tag{3.137}$$

which in turn gives (3.136b) as

$$A_t + 2AA_x - A_{xx} + g_x = 0. \tag{3.138}$$

The Hopf-Cole transformation (3.99) i.e.

$$A = -\phi_x/\phi \tag{3.139}$$

works well here, where substitution and integrating (3.138) gives

$$\phi_t - \phi_{xx} + g(t, x)\phi = 0. \tag{3.140}$$

Note that the function of integration is omitted without loss of generality. Combining the results, solutions of
$$u_t = u_{xx} + (g(t, x) + 2(\ln \phi)_{xx})u \tag{3.141}$$

are obtained via the Darboux transformation

$$u = v_x - \frac{\phi_x}{\phi}v, \tag{3.142}$$

where v and ϕ satisfy (3.133) and (3.140), respectively.

3.5.2.2 Second Order Darboux Transformations

We now extend the Darboux transformation to second order and seek solutions of

$$u_t = u_{xx} + f(t, x)u \tag{3.143}$$

via the Darboux transformations of the form

$$u = v_{xx} + A(t, x)v_x + B(t, x)v, \tag{3.144}$$

where v satisfies (3.133). As we did previously, substituting (3.144) into (3.143), imposing (3.133) and isolating coefficients with respect to v, v_x, and v_{xx}, gives

$$2A_x + f - g = 0, \tag{3.145a}$$
$$A_t - A_{xx} - 2B_x - fA + 2g_x + gA = 0, \tag{3.145b}$$
$$B_t - B_{xx} + gB - fB + Ag_x + g_{xx} = 0. \tag{3.145c}$$

From (3.145a) we again see that

$$f = g - 2A_x, \tag{3.146}$$

which, in turn, gives (3.145b) and (3.145c) as

$$A_t + 2AA_x - A_{xx} - 2B_x + 2g_x = 0. \tag{3.147a}$$
$$B_t + 2BA_x - B_{xx} + g_{xx} + Ag_x = 0. \tag{3.147b}$$

If we introduce the matrices

$$\Omega = \begin{bmatrix} 0 & -1 \\ B & A \end{bmatrix}, \quad G = \begin{bmatrix} g & 0 \\ g_x & g \end{bmatrix}, \tag{3.148}$$

the system of equations (3.147) conveniently becomes

$$\Omega_t + 2\Omega_x\Omega - \Omega_{xx} + \Omega G - G\Omega + G_x = 0. \tag{3.149}$$

Again, the Matrix Hopf-Cole transformation

$$\Omega = -\Phi_x\Phi^{-1} \tag{3.150}$$

works well here, leading to the linear Matrix equation

$$\Phi_t - \Phi_{xx} - G\Phi = 0. \tag{3.151}$$

As the process is now identical to that presented in (3.117)–(3.124), we simply state our results. Solutions of

$$u_t = u_{xx} + (g(t, x) + 2(\ln \phi)_{xx})\, u \tag{3.152}$$

can be obtained via the Darboux transformation

$$u = v_{xx} - \frac{\begin{vmatrix} \omega_1 & \omega_2 \\ \omega_{1xx} & \omega_{2xx} \end{vmatrix}}{\begin{vmatrix} \omega_1 & \omega_2 \\ \omega_{1x} & \omega_{2x} \end{vmatrix}}\, v_x \; + \; \frac{\begin{vmatrix} \omega_{1x} & \omega_{2x} \\ \omega_{1xx} & \omega_{2xx} \end{vmatrix}}{\begin{vmatrix} \omega_1 & \omega_2 \\ \omega_{1x} & \omega_{2x} \end{vmatrix}}\, v, \tag{3.153}$$

where $\phi = \omega_1 \omega_2' - \omega_2 \omega_1'$, ω_1, ω_2, and v are all independent solutions of

$$w_t = w_{xx} + g(t, x)w. \tag{3.154}$$

These results also generalize to nth order Darboux transformations; we refer the reader to Arrigo and Hickling [12] for further details.

3.5.3 Darboux Transformations Between Two Wave Equations

In this section we wish to connect solutions to two different wave equations. In particular, find wave equations of the form

$$u_{tt} = u_{xx} + f(x)u, \tag{3.155}$$

that admit Darboux transformations where v satisfies

$$v_{tt} = v_{xx}. \tag{3.156}$$

Again, we will consider first and second order transformations

3.5.3.1 First Order Darboux Transformations

We will start with first order Darboux transformations as we did earlier and seek solutions of (3.155) via the Darboux transformations of the form

$$u = v_x + A(x)v, \tag{3.157}$$

where v satisfies (3.156). As we did previously, substituting (3.157) into (3.155), imposing (3.156) and isolating coefficients with respect to v and v_x, gives

$$2A' + f = 0, \tag{3.158a}$$

$$A'' + fA = 0, \tag{3.158b}$$

where prime denotes differentiation with respect to the argument. From (3.158a) we see that

$$f = -2A',$$
(3.159)

which in turn gives (3.158b) as

$$A'' - 2AA' = 0.$$
(3.160)

This can be integrated once giving

$$A' - A^2 = c,$$
(3.161)

where c is a constant of integration. If we introduce a Hopf-Cole transformation

$$A = -\phi'/\phi,$$
(3.162)

where $\phi = \phi(x)$; then (3.161) becomes

$$\phi'' + c\phi = 0.$$
(3.163)

Combining the results, solutions of

$$u_{tt} = u_{xx} + 2(\ln \phi)'' u$$
(3.164)

are obtained via the Darboux transformation

$$u = v_x - \frac{\phi'}{\phi} v,$$
(3.165)

where v is a solution of the wave equation (3.156) and ϕ a solution of (3.163).

3.5.3.2 Second Order Darboux Transformations

We now extend the Darboux transformation to second order and seek solutions of

$$u_t = u_{xx} + f(x)u$$
(3.166)

via the Darboux transformations of the form

$$u = v_{xx} + A(x)v_x + B(x)v,$$
(3.167)

where v satisfies (3.156). As we did previously, substituting (3.167) into (3.166), imposing (3.156) and isolating coefficients with respect to v, v_x, and v_{xx}, gives

$$2A' + f = 0,$$
(3.168a)

$$A'' + 2B' + fA = 0.$$
(3.168b)

$$B'' + fB = 0.$$
(3.168c)

From (3.168a) we again see that

$$f = -2A', \tag{3.169}$$

which, in turn, gives (3.168b) and (3.168c) as

$$A'' - 2AA' + 2B' = 0, \tag{3.170a}$$
$$B'' - 2BA' = 0. \tag{3.170b}$$

If we introduce the matrix

$$\Omega = \begin{bmatrix} 0 & -1 \\ B & A \end{bmatrix}, \tag{3.171}$$

the system of equations (3.170) conveniently becomes

$$\Omega'' - 2\Omega'\Omega = 0. \tag{3.172}$$

Again, the Matrix Hopf-Cole transformation

$$\Omega = -\Phi'\Phi^{-1} \tag{3.173}$$

works well here leading to the linear Matrix equation

$$\left(\Phi^{-1}\Phi_{xx}\right)' = 0, \tag{3.174}$$

which we integrate giving

$$\Phi_{xx} = \Phi \mathbf{c}, \tag{3.175}$$

where $\mathbf{c}$ is a 2 by 2 matrix of arbitrary constants, namely

$$\mathbf{c} = \begin{bmatrix} c_{11} & c_{12} \\ c_{21} & c_{22} \end{bmatrix}. \tag{3.176}$$

As the process is now identical to that presented in (3.117)–(3.124), we simply state our results. Solutions of

$$u_{tt} = u_{xx} + 2(\ln \phi)'' u \tag{3.177}$$

are obtained via the Darboux transformation

$$u = v_{xx} - \frac{\begin{vmatrix} \omega_1 & \omega_2 \\ \omega_1'' & \omega_2'' \end{vmatrix}}{\begin{vmatrix} \omega_1 & \omega_2 \\ \omega_1' & \omega_2' \end{vmatrix}} v_x + \frac{\begin{vmatrix} \omega_1' & \omega_2' \\ \omega_1' & \omega_2' \end{vmatrix}}{\begin{vmatrix} \omega_1 & \omega_2 \\ \omega_1' & \omega_2' \end{vmatrix}} v, \tag{3.178}$$

where $\phi = \omega_1\omega_2' - \omega_2\omega_1'$, with ω_1, ω_2 and v satisfying the ODEs

$$\omega_1'' = c_{11}\omega_1 + c_{21}\omega_2,$$
$$\omega_2'' = c_{12}\omega_1 + c_{22}\omega_2,$$
(3.179)

and the wave equation (3.156).

3.6   Exercises

1. Find a Darboux transformation in the form $y = z' + A(x)z$ connecting

$$y'' + f(x)y = 0, \quad z'' + g(x)z = 0.$$

2. Darboux transformations for (3.93) naturally extend to nth order where the nth order Darboux transformation is given by

$$u = \frac{W(\omega_1, \omega_2, \ldots, \omega_n, v)}{W(\omega_1, \omega_2, \ldots, \omega_n)}$$
(3.180)

applicable for
$$f = 2(\ln W)_{xx} \quad \text{where} \quad W = W(\omega_1, \omega_2, \ldots, \omega_n).$$
(3.181)

(a) Show that by picking the following polynomial solutions of the heat equation

$$\omega_1 = x,$$
$$\omega_2 = x^3 + 6xt,$$
$$\omega_3 = x^5 + 20x^3 t + 60xt^2,$$
$$\omega_4 = x^7 + 42x^5 t + 420x^3 t^2 + 840xt^3,$$
$$\omega_5 = x^9 + 72x^7 t + 1512x^5 t^2 + 10080x^3 t^3 + 15120xt^4,$$

that f in the case of $n = 3, 4$ and 5, becomes

$$f = -\frac{12}{x^2}, \quad f = -\frac{20}{x^2}, \quad f = -\frac{30}{x^2}.$$
(3.182)

Conjecture the general form of f is the case of arbitrary n.

(b) Show that by picking the following solutions of the heat equation

$$\omega_n = \begin{cases} e^{n^2 t} \cosh nx & \text{for } n \text{ odd} \\ e^{n^2 t} \sinh nx & \text{for } n \text{ even} \end{cases}$$
(3.183)

that for $n = 3, 4$ and 5, that f becomes

$$f = \frac{12}{\cosh^2 x}, \quad f = \frac{20}{\cosh^2 x}, \quad f = \frac{30}{\cosh^2 x}.$$
(3.184)

Conjecture the general form of f is the case of arbitrary n.

3. Show by a suitable transformation $u = F(v)$, the Thomas equation

$$u_{xy} + au_x u_y + bu_x + cu_y = 0$$

(a, b, and c constant) can be linearized.

4. Show that the forced Burgers' equation

$$u_t + 2uu_x = u_{xx} + f(x)$$

can be linearized via the Hopf-Cole transformation.

References

1. J.M. Burgers, A mathematical model illustrating the theory of turbulence. Adv. App. Mech. **1**, 171–199 (1948)
2. H. Bateman, Some recent researches on the motion of fluids. Mon. Weather Rev. **43**(4), 163–170 (1915)
3. E. Hopf, The partial differential equation $u_t + uu_x = \mu u_{xx}$. Commun. Pure Appl. Math. **3**, 201–230 (1950)
4. J.D. Cole, On a quasi-linear parabolic equation occurring in aerodynamics. Quart. Appl. Math. **9**, 225–236 (1951)
5. A.R. Forsyth, *Theory of Differential Equations,* Vol. VI (Cambridge University Press, 1906) P. 102, Ex. 3
6. D.J. Korteweg, F. de Vries, On the change of form of long waves advancing in a rectangular canal, and on a new type of long stationary waves. Philos. Mag. **39**, 422–443 (1895)
7. R.M. Miura, The Korteweg−deVries equation: a survey of results. SIAM Rev. **18**(3), 412–459 (1978)
8. R.M. Miura, Korteweg-de Vries equation and generalizations. I. A remarkable explicit nonlinear transformation. J. Math. Phys. **9**, 1202–1204 (1968)
9. R.M. Miura, C.S. Gardner, M.D. Kruskal, Korteweg-de Vries equation and generalizations. II. Existence of conservation laws and constants of motion. J. Math. Phys. **9**, 1204–1209 (1968)
10. G.M. Cox, J.M. Hill, N. Thamwattana, A formal exact mathematical solution for a sloping rat-hole in a highly frictional granular solid. Acta. Mech. **170**, 127–147 (2004)
11. D.J. Arrigo, F. Hickling, A Darboux transformation for a class of linear parabolic partial differential equation. J. Phys. A: Math. Gen. **35**, 389–399 (2002)
12. D.J. Arrigo, F. Hickling, An nth order Darboux transformation for the one−dimensional time dependent Schrodinger equation. J. Phys. A: Math. Gen. **36**, 1615–1621 (2003)

Point and Contact Transformations

In an introductory course in PDEs we found that by introducing new variables

$$x = f(r,s), \quad y = g(r,s), \quad \frac{\partial(x,y)}{\partial(r,s)} \neq 0, \tag{4.1}$$

we were able to transform PDEs of the form

$$Au_{xx} + Bu_{xy} + Cu_{yy} + \text{l.o.t.s} = 0, \tag{4.2}$$

where A, B and C are functions of (x,y) and l.o.t.s are lower order terms to standard form, i.e.

$$\begin{aligned} u_{ss} + \text{l.o.t.s} &= 0 \quad \text{parabolic,} \\ u_{rs} + \text{l.o.t.s} &= 0 \quad \text{modified hyperbolic,} \\ u_{rr} + u_{ss} + \text{l.o.t.s} &= 0 \quad \text{elliptic.} \end{aligned} \tag{4.3}$$

We now ask, can we generalize these transformation? For example, can we consider transformations of the form

$$x = F(X,Y,U), \quad y = G(X,Y,U), \quad u = H(X,Y,U), \tag{4.4}$$

where $U = U(X,Y)$? Transformations of this type are referred to as *point transformations*. Consider, for example the transformation

$$x = X, \quad y = U, \quad u = Y. \tag{4.5}$$

In order to see the effect on a PDE, we will need to calculate how derivatives transform. The easiest way is using Jacobians. For example, the derivative u_x transforms as

© The Author(s), under exclusive license to Springer Nature Switzerland AG 2022

D. Arrigo, *Analytical Methods for Solving Nonlinear Partial Differential Equations*,

Synthesis Lectures on Mathematics & Statistics.

https://doi.org/10.1007/978-3-031-17069-0_4

$$
u_x = \frac{\partial(u, y)}{\partial(x, y)} = \frac{\dfrac{\partial(u, y)}{\partial(X, Y)}}{\dfrac{\partial(x, y)}{\partial(X, Y)}} = \frac{\dfrac{\partial(Y, U)}{\partial(X, Y)}}{\dfrac{\partial(X, U)}{\partial(X, Y)}} = -\frac{U_X}{U_Y}, \tag{4.6}
$$

noting that we have used the transformation (4.5) in (4.6). Similarly for u_y.

$$
u_y = \frac{\partial(x, u)}{\partial(x, y)} = \frac{\dfrac{\partial(x, u)}{\partial(X, Y)}}{\dfrac{\partial(x, y)}{\partial(X, Y)}} = \frac{\dfrac{\partial(X, Y)}{\partial(X, Y)}}{\dfrac{\partial(X, U)}{\partial(X, Y)}} = \frac{1}{U_Y}. \tag{4.7}
$$

The process easily extends to second order derivatives. For example, u_{xx} transforms as

$$
u_{xx} = \frac{\partial(u_x, y)}{\partial(x, y)} = \frac{\dfrac{\partial(u_x, y)}{\partial(X, Y)}}{\dfrac{\partial(x, y)}{\partial(X, Y)}} = \frac{\dfrac{\partial(-\frac{U_Y}{U_X}, U)}{\partial(X, Y)}}{\dfrac{\partial(X, U)}{\partial(X, Y)}}
$$

$$
= -\frac{U_Y^2 U_{XX} - 2U_X U_Y U_{XY} + U_X^2 U_{YY}}{U_Y^3}, \tag{4.8}
$$

while u_{xy} and u_{yy} transform as

$$
u_{xy} = \frac{\partial(u_y, y)}{\partial(x, y)} = \frac{\dfrac{\partial(u_y, y)}{\partial(X, Y)}}{\dfrac{\partial(x, y)}{\partial(X, Y)}} = \frac{\dfrac{\partial(\frac{1}{U_Y}, U)}{\partial(X, Y)}}{\dfrac{\partial(X, U)}{\partial(X, Y)}}
$$

$$
= \frac{U_X U_{YY} - U_Y U_{XY}}{U_Y^3}, \tag{4.9}
$$

and

$$
u_{yy} = \frac{\partial(x, u_y)}{\partial(x, y)} = \frac{\dfrac{\partial(x, u_y)}{\partial(X, Y)}}{\dfrac{\partial(x, y)}{\partial(X, Y)}} = \frac{\dfrac{\partial(X, \frac{1}{U_Y})}{\partial(X, Y)}}{\dfrac{\partial(X, U)}{\partial(X, Y)}}
$$

$$
= -\frac{U_{YY}}{U_Y^3}. \tag{4.10}
$$

Thus, under the transformation (4.5), the first and second order derivatives transform as

$$u_x = -\frac{U_X}{U_Y}, \quad u_y = \frac{1}{U_Y}, \tag{4.11}$$

$$u_{xx} = -\frac{U_Y^2 U_{XX} - 2U_X U_Y U_{XY} + U_X^2 U_{YY}}{U_Y^3}, \quad u_{xy} = \frac{U_X U_{YY} - U_Y U_{XY}}{U_Y^3}, \quad u_{yy} = -\frac{U_{YY}}{U_Y^3}.$$

We now ask - why would this be useful? Consider, for example, the heat equation

$$u_{xx} = u_y. \tag{4.12}$$

Under (4.5) (using the derivatives obtained in (4.11)), (4.12) would become

$$-\frac{U_Y^2 U_{XX} - 2U_X U_Y U_{XY} + U_X^2 U_{YY}}{U_Y^3} = \frac{1}{U_Y},$$

or

$$U_Y^2 U_{XX} - 2U_X U_Y U_{XY} + U_X^2 U_{YY} + U_Y^2 = 0.$$

So one may ask—what was the point? We clearly made the problem more difficult. However, suppose we started with the nonlinear PDE

$$u_y^2 u_{xx} - 2u_x u_y u_{xy} + u_x^2 u_{yy} + u_y^2 = 0, \tag{4.13}$$

a PDE given in (3.67) (with $c(v) = 1$), then under (4.5), (4.13) becomes

$$U_{XX} = U_Y,$$

the linear heat equation!

4.1   Contact Transformations

We now extend transformations to include first derivatives

$$x = F(X, Y, U, P, Q), \quad y = G(X, Y, U, P, Q), \quad u = H(X, Y, U, P, Q), \tag{4.14}$$

where $U = U(X, Y)$, $P = U_X$ and $Q = U_Y$. We also need the relation that

$$u_x = M(X, Y, U, P, Q), \quad u_y = N(X, Y, U, P, Q), \tag{4.15}$$

and impose the condition that (4.15) cannot have higher order derivatives. Before talking about conditions which will ensure this, we consider three very famous contact transformations:

1. the Hodograph transformation,
2. the Legendre transformation, and
3. the Ampere transformation.

4.1.1   Hodograph Transformation

These transformations are of the form

$$x = X, \quad y = U, \quad u = Y, \tag{4.16}$$

or

$$x = U, \quad y = Y, \quad u = X. \tag{4.17}$$

The first transformation (4.16) was previously given in (4.5). For the second transformation (4.17), first order derivatives transform as

$$u_x = \frac{1}{U_X}, \quad u_y = -\frac{U_Y}{U_X}, \tag{4.18}$$

and the second order derivatives transform as

$$
\begin{aligned}
u_{xx} &= -\frac{U_{XX}}{U_X^3}, \\
u_{xy} &= \frac{U_Y U_{XX} - U_X U_{XY}}{U_X^3}, \\
u_{yy} &= -\frac{U_Y^2 U_{XX} - 2U_X U_Y U_{XY} + U_X^2 U_{YY}}{U_X^3}.
\end{aligned}
\tag{4.19}
$$

4.1.2   Legendre Transformation

These transformations are of the form

$$x = U_X, \quad y = U_Y, \quad u = X U_X + Y U_Y - U. \tag{4.20}$$

Using Jacobians as we did for point transformations, we find the first derivatives transform as

$$u_x = X, \quad u_y = Y, \tag{4.21}$$

and the second derivatives transform as

$$u_{xx} = \frac{U_{YY}}{U_{XX} U_{YY} - U_{XY}^2}, \quad u_{xy} = -\frac{U_{XY}}{U_{XX} U_{YY} - U_{XY}^2}, \quad u_{yy} = \frac{U_{XX}}{U_{XX} U_{YY} - U_{XY}^2}. \tag{4.22}$$

4.1.3 Ampere Transformation

These transformations are of the form

$$x = U_X, \quad y = Y, \quad u = XU_X - U. \tag{4.23}$$

Using Jacobians, we find the first derivatives transform as

$$u_x = X, \quad u_y = -U_Y, \tag{4.24}$$

and the second derivatives transform as

$$u_{xx} = \frac{1}{U_{XX}}, \quad u_{xy} = -\frac{U_{XY}}{U_{XX}}, \quad u_{yy} = -\frac{U_{XX}U_{YY} - U_{XY}^2}{U_{XX}}. \tag{4.25}$$

Note that we could easily have chosen

$$x = X, \quad y = U_Y, \quad u = YU_Y - U, \tag{4.26}$$

and the derivatives would have transformed similarly.

We now consider some examples.

Example 4.1 Consider

$$u_y^2 u_{xx} - 2u_x u_y u_{xy} + u_x^2 u_{yy} + u_y^3 = 0. \tag{4.27}$$

Under the Hodograph transformation (4.16), Eq. (4.27) becomes

$$U_{XX} - 1 = 0, \tag{4.28}$$

which integrates giving

$$U = \frac{1}{2}X^2 + F(Y)X + G(Y), \tag{4.29}$$

where F and G are arbitrary functions; via (4.16) we obtain the exact solution

$$y = \frac{1}{2}x^2 + F(u)x + G(u). \tag{4.30}$$

Example 4.2 Consider the nonlinear diffusion equation

$$u_t = \frac{u_{xx}}{u_x^2}. \tag{4.31}$$

Under the Hodograph transformation

$$t = T, \quad x = U, \quad u = X, \tag{4.32}$$

Equation (4.31) becomes

$$U_T = U_{XX}, \tag{4.33}$$

the heat equation!

Example 4.3 Consider

$$u_{xx} - x\left(u_{xx}u_{yy} - u_{xy}^2\right) = 0. \tag{4.34}$$

Under the Legendre transformation (4.20), Eq. (4.34) becomes

$$\frac{U_{YY}}{U_{XX}U_{YY} - U_{XY}^2} - U_X\left(\frac{U_{YY}}{U_{XX}U_{YY} - U_{XY}^2}\frac{U_{XX}}{U_{XX}U_{YY} - U_{XY}^2} - \frac{U_{XY}^2}{(U_{XX}U_{YY} - U_{XY}^2)^2}\right) = 0, \tag{4.35}$$

which, after simplification, becomes

$$U_{YY} - U_X = 0, \tag{4.36}$$

the heat equation.

Example 4.4 Consider

$$u_{xx} - u_{yy} - \left(u_x - u_y\right)\left(u_{xx}u_{yy} - u_{xy}^2\right) = 0. \tag{4.37}$$

Under the Legendre transformation (4.20), Eq. (4.37) becomes

$$\frac{U_{YY} - U_{XX}}{U_{XX}U_{YY} - U_{XY}^2} - (X - Y)\left(\frac{U_{YY}U_{XX} - U_{XY}^2}{(U_{XX}U_{YY} - U_{XY}^2)^2}\right) = 0, \tag{4.38}$$

which, after simplification becomes

$$U_{XX} - U_{YY} = X - Y. \tag{4.39}$$

A particular solution of (4.39) is $U = (X^3 + Y^3)/6$; with this, we can transform (4.39) to the standard wave equation

$$V_{XX} - V_{YY} = 0, \tag{4.40}$$

via

$$U = V + \frac{1}{6}X^3 + \frac{1}{6}Y^3. \tag{4.41}$$

The solution of (4.40) is

$$V = F(X + Y) + G(X - Y). \tag{4.42}$$

Composing (4.42), (4.41) and (4.20) gives

$$x = F'(X+Y) + G'(X-Y) + \frac{1}{2}X^2,$$

$$y = F'(X+Y) - G'(X-Y) + \frac{1}{2}Y^2, \tag{4.43}$$

$$u = (X+Y)F'(X+Y) + (X-Y)G'(X-Y)$$
$$- F(X+Y) - G(X+Y) + \frac{1}{3}X^3 + \frac{1}{3}Y^3,$$

the exact solution of (4.37)

Example 4.5 Consider

$$u_{xx}u_{yy} - u_{xy}^2 = 1. \tag{4.44}$$

Under the Ampere transformation (4.23), (4.44) becomes

$$-\frac{1}{U_{XX}}\frac{U_{XX}U_{YY} - U_{XY}^2}{U_{XX}} - \frac{U_{XY}^2}{U_{XX}^2} = 1, \tag{4.45}$$

which, after simplification, becomes

$$U_{XX} + U_{YY} = 0, \tag{4.46}$$

Laplace's equation!

It is possible to combine several transformations as the following example demonstrates.

Example 4.6 We consider the PDE

$$u_{xx}u_{yy} - u_{xy}^2 = \left(u_x^2 + u_y^2\right)^2. \tag{4.47}$$

This PDE appears in elasticity [2]. Under the Ampere transformation

$$x = X, \quad y = U_Y, \quad u = YU_Y - U, \quad u_x = -U_X, \quad u_y = Y, \tag{4.48}$$

(4.47) becomes
$$U_{XX} + \left(U_X^2 + Y^2\right)^2 U_{YY} = 0. \tag{4.49}$$

If we let $U = YV$, (4.49) becomes

$$V_{XX} + Y^3 \left(V_X^2 + 1\right)^2 (YV_{YY} + 2V_Y) = 0. \tag{4.50}$$

Introducing the new variable $Y = 1/S$, the Y derivatives transform as

$$V_Y = -S^2 V_S, \quad V_{YY} = S^4 V_{SS} + 2S^3 V_S,$$

so that (4.50) becomes

$$V_{XX} + \left(V_X^2 + 1\right)^2 V_{SS} = 0. \tag{4.51}$$

The transformation (4.48) under the change of variables so far is

$$x = X, \quad y = V - S V_S, \quad u = -V_S, \quad u_x = -\frac{V_S}{S}, \quad u_y = \frac{1}{S}. \tag{4.52}$$

Next, we perform a Legendre transformation on (4.51)

$$X = W_\xi, \quad S = W_\eta, \quad V = \xi W_\xi + \eta W_\eta - W, \quad V_X = \xi, \quad V_S = \eta, \tag{4.53}$$

giving

$$W_{\eta\eta} + \left(\xi^2 + 1\right)^2 W_{\xi\xi} = 0. \tag{4.54}$$

If we let

$$\xi = \tan \gamma, \quad W = Q \sec \gamma, \tag{4.55}$$

then (4.54) becomes

$$Q_{\gamma\gamma} + Q_{\eta\eta} + Q = 0. \tag{4.56}$$

Composing all the transformations into one (with new variables) gives

$$x = \cos X U_X + \sin X U, \quad y = \sin X U_X - \cos X U, \quad u = Y, \tag{4.57}$$

and under (4.57), (4.47) is transformed to the Helmholtz equation

$$U_{XX} + U_{YY} + U = 0. \tag{4.58}$$

These results are presented in Arrigo and Hill [2].

It is important to realize that not all transformations of the form (4.14) are contact transformations. Consider, for example,

$$x = X, \quad y = U, \quad u = U_X. \tag{4.59}$$

First derivatives transform as

$$u_x = \frac{U_Y U_{XX} - U_X U_{XY}}{U_Y}, \quad u_y = \frac{U_{XY}}{U_Y}, \tag{4.60}$$

showing that they possess second order derivative terms. Thus, we wish to establish conditions to guarantee that a transformation is, in fact, a contact transformation.

4.2   Contact Condition

We first consider the problem with ODEs before moving to PDEs. Suppose we have a transformation

$$X = X(x, y, p), \quad Y = Y(x, y, p), \quad P = P(x, y, p), \tag{4.61}$$

where $p = \dfrac{dy}{dx}$ and $P = \dfrac{dY}{dX}$. We wish to calculate $\dfrac{dY}{dX}$. From (4.61) we have

$$\frac{dY}{dX} = \frac{Y_x + Y_y \dfrac{dy}{dx} + Y_p \dfrac{d^2 y}{dx^2}}{X_x + X_y \dfrac{dy}{dx} + X_p \dfrac{d^2 y}{dx^2}} = P, \tag{4.62}$$

or

$$Y_x + Y_y \frac{dy}{dx} + Y_p \frac{d^2 y}{dx^2} = P \left(X_x + X_y \frac{dy}{dx} + X_p \frac{d^2 y}{dx^2} \right). \tag{4.63}$$

As X and Y are independent of $\dfrac{d^2 y}{dx^2}$, then from (4.63) we have

$$Y_x + pY_y = P \left(X_x + pX_y \right), \tag{4.64a}$$
$$Y_p = PX_p. \tag{4.64b}$$

The set of Eq. (4.64) is know as the contact conditions. In 1872 Lie [1] gave the following definition of a contact transformation: if X, Y, P are independent functions of x, y, p such that

$$dY - PdX = \lambda(dy - pdx) \tag{4.65}$$

for some $\lambda = \lambda(x, y, p)$, then $X = X(x, y, p), \quad Y = Y(x, y, p), \quad P = P(x, y, p)$ is a contact transformation. Inserting the appropriate differentials into (4.65) gives

$$Y_x dx + Y_y dy + Y_p dp - P(X_x dx + X_y dy + X_p dp) = \lambda(dy - pdx). \tag{4.66}$$

Isolating coefficients with respect to dx, dy and dp gives

$$Y_x - PX_x = -p\lambda, \tag{4.67a}$$
$$Y_y - PX_y = \lambda, \tag{4.67b}$$
$$Y_p - PX_p = 0, \tag{4.67c}$$

and it is a simple matter to show that eliminating λ in (4.67) gives (4.64). With Lie's definition in hand, we can extend this to PDEs.

If

$$dU - PdX - QdY = \lambda (du - pdx - qdy), \tag{4.68}$$

then under the transformations

$$X = X(x, y, u, p, q), \quad Y = Y(x, y, u, p, q), \quad U = U(x, y, u, p, q), \tag{4.69}$$

Equation (4.68) becomes

$$U_x \, dx + U_y \, dy + U_u \, du + U_p \, dp + U_q \, dq \tag{4.70}$$
$$-P \left(X_x \, dx + X_y \, dy + X_u \, du + X_p \, dp + X_q \, dq \right) \tag{4.71}$$
$$-Q \left(Y_x \, dx + Y_y \, dy + Y_u \, du + Y_p \, dp + Y_q \, dq \right) \tag{4.72}$$
$$= \lambda \left(dU - U_X dX - U_Y dY \right). \tag{4.73}$$

Expanding and equating differentials gives

$$\begin{aligned}
U_p - PX_p - QY_p &= 0, \\
U_q - PX_q - QY_q &= 0, \\
U_u - PX_u - QY_u &= \lambda, \\
U_x - PX_x - QY_x &= -\lambda p, \\
U_y - PX_y - QY_y &= -\lambda q.
\end{aligned} \tag{4.74}$$

The relations (4.74) are known as the *contact conditions*.

Example 4.7 Consider the Hodograph transformation

$$X = x, \quad Y = u \quad U = y. \tag{4.75}$$

Calculating the first derivatives, we find

$$P = -\frac{p}{q}, \quad Q = \frac{1}{q}. \tag{4.76}$$

Substituting (4.75) and (4.76) into (4.74) show these are satisfied if $\lambda = -1$.

Example 4.8 Consider the Legendre transformation

$$X = p, \quad Y = q, \quad U = xp + yq - u. \tag{4.77}$$

Calculating the first derivatives, we find

$$P = X, \quad Q = Y. \tag{4.78}$$

Substituting (4.77) and (4.78) into (4.74) show these are satisfied if $\lambda = -1$.

Example 4.9 Consider the Ampere transformation

$$X = x, \quad Y = q, \quad U = yq - u. \tag{4.79}$$

Calculating the first derivatives, we find

$$P = -p, \quad Q = y. \tag{4.80}$$

Substituting (4.79) and (4.80) into (4.74) shows these are satisfied if $\lambda = -1$.

Example 4.10 Consider

$$X = x + q, \quad Y = y + p, \quad U = u + pq. \tag{4.81}$$

Calculating the first derivatives we find

$$P = p, \quad Q = q. \tag{4.82}$$

Substituting (4.81) and (4.82) into (4.74) show these are satisfied if $\lambda = 1$; thus, (4.81) is a contact transformation.

4.3 Plateau Problem

Consider the surface area of $u = u(x, y)$ on some region $\mathbb{R}$

$$SA = \iint_{\mathbb{R}} \sqrt{1 + u_x^2 + u_y^2}\, dA. \tag{4.83}$$

For this surface to be a minimum, then the Euler-Lagrange equation from the calculus of variations is

$$\frac{\partial}{\partial x}\left(\frac{\partial L}{\partial u_x}\right) + \frac{\partial}{\partial y}\left(\frac{\partial L}{\partial u_y}\right) - \frac{\partial L}{\partial u} = 0 \tag{4.84}$$

with $L = \sqrt{1 + u_x^2 + u_y^2}$. This gives

$$\left(1 + u_y^2\right) u_{xx} - 2u_x u_y u_{xy} + \left(1 + u_x^2\right) u_{yy} = 0. \tag{4.85}$$

It is well known that solutions of (4.85) can be generated by the Enneper-Weierstrass formulas given by

$$x = Re \int f(1 - g^2)dz,$$

$$y = Re \int if(1 + g^2)dz, \tag{4.86}$$

$$u = Re \int 2f g \, dz.$$

For example, choosing $f = 1$ and $g = z$ gives

$$x = Re\left(z - \frac{z^3}{3}\right),$$

$$y = Re \, i\left(z + \frac{z^3}{3}\right), \tag{4.87}$$

$$u = Re \, z^2,$$

and denoting $z = p + iq$ gives rise to

$$x = p - \frac{p^3}{3} + pq^2,$$

$$y = -q - p^2q + \frac{q^3}{3}, \tag{4.88}$$

$$u = p^2 - q^2,$$

commonly referred to as Enneper's minimal surface. Here we present a new way of generating a solution to (4.85). We will show that (4.85) is linearizable and furthermore, we can represent its solution parametrically in terms of solutions of Laplace's equation.

4.3.1   Linearization

Under the Legendre transformation

$$x = U_X, \quad y = U_Y \quad u = XU_X + YU_Y - U, \tag{4.89}$$

first order derivatives transform as

$$u_x = X, \quad u_y = Y, \tag{4.90}$$

while second order derivatives transform as

$$u_{xx} = \frac{U_{YY}}{U_{XX}U_{YY} - U_{XY}^2}, \quad u_{xy} = -\frac{U_{XY}}{U_{XX}U_{YY} - U_{XY}^2}, \quad u_{yy} = \frac{U_{YY}}{U_{XX}U_{YY} - U_{XY}^2}, \tag{4.91}$$

thus transforming (4.85) to the linear PDE

$$\left(1 + X^2\right) U_{XX} + 2XY U_{XY} + \left(1 + Y^2\right) U_{YYs} = 0. \tag{4.92}$$

If we denote $A = 1 + X^2$, $B = 2XY$ and $C = 1 + Y^2$ then $B^2 - 4AC < 0$ so that the PDE (4.92) is elliptic. We now show that (4.92) is transformable to Laplace's equation. Introducing polar coordinates

$$X = r \cos s, \quad Y = r \sin s \tag{4.93}$$

transforms (4.92) to

$$\left(r^4 + r^2\right) U_{rr} + U_{ss} + r U_r = 0. \tag{4.94}$$

Further, if we let

$$r = \frac{1}{\sinh a}, \quad s = b \tag{4.95}$$

then (4.94) becomes

$$U_{aa} + U_{bb} + \frac{2U_a}{\sinh a \cosh a} = 0. \tag{4.96}$$

Next we transform the dependent variable U as

$$U = \frac{V}{\tanh a}, \tag{4.97}$$

leading to

$$V_{aa} + V_{bb} + \frac{2V}{\cosh^2 a} = 0. \tag{4.98}$$

Finally, if we introduce the first order Darboux transformation

$$V = W_a - \tanh a \, W, \tag{4.99}$$

then Eq. (4.98) becomes

$$\frac{\partial}{\partial a} \left(\nabla^2 W\right) - \tanh a \nabla^2 W = 0, \tag{4.100}$$

where $\nabla^2 W = 0$ is Laplace's equation. Integrating (4.100) gives

$$\nabla^2 W = F(b) \cosh a, \tag{4.101}$$

where $F(b)$ is arbitrary, but we can set $F(b) = 0$ without loss of generality. The rationale is as follows. If $F(b)$ is arbitrary, then we can set $F = G'' + G$ (G is arbitrary) so

$$\nabla^2 W = \left(G''(b) + G(b)\right) \cosh a. \tag{4.102}$$

If

$$W = \widetilde{W} + G(b) \cosh a, \tag{4.103}$$

then (4.102) gives rise to Laplace's equation

$$\nabla^2 \widetilde{W} = 0. \tag{4.104}$$

Further substitution into (4.99) shows it is left unchanged.

In order to connect solutions of the original equation (4.85) to Laplace's equation, it is necessary to compose the transformations (4.89), (4.93), (4.95), (4.97) and (4.99). This gives rise to the following result:

Exact solutions to the minimal surface equation

$$\left(1 + u_y^2\right) u_{xx} - 2u_x u_y u_{xy} + \left(1 + u_x^2\right) u_{yy} = 0 \tag{4.105}$$

are given parametrically by

$$x = \cosh a \cos b \; W_a + \sinh a \sin b \; W_b - \sinh a \cos b \; W_{aa} - \cosh a \sin b \; W_{ab},$$
$$y = \cosh a \sin b \; W_a - \sinh a \cos b \; W_b + \cosh a \cos b \; W_{ab} + \sinh a \sin b \; W_{bb}, \tag{4.106}$$
$$u = W - W_{aa},$$

where W satisfies Laplace's equation

$$W_{aa} + W_{bb} = 0. \tag{4.107}$$

In the next section we will show that several well-known minimal surfaces may be obtained with simple solutions of Laplace's equation.

4.3.2   Well Known Minimal Surfaces

4.3.2.1 Catenoid and Helicoid

Probably the simplest non-zero solution to Laplace's equation, (4.107) is given by

$$W = a \sin \alpha + b \cos \alpha, \tag{4.108}$$

where α is constant. Substituting this into (4.106) gives rise to

$$x = \cosh a \cos b \; \sin \alpha + \sinh a \sin b \; \cos \alpha,$$
$$y = \cosh a \sin b \; \sin \alpha - \sinh a \cos b \; \cos \alpha, \tag{4.109}$$
$$u = a \sin \alpha + b \cos \alpha.$$

Setting $\alpha = 0$ gives

$$x = \sinh a \sin b,$$
$$y = - \sinh a \cos b, \tag{4.110}$$
$$u = b,$$

the surface of a helicoid, while setting $\alpha = \frac{\pi}{2}$ gives

$$x = \cosh a \cos b,$$
$$y = \cosh a \sin b, \tag{4.111}$$
$$u = a,$$

the surface of a catenoid [6].

4.3.2.2 Enneper's Surface

By choosing

$$W = -\frac{1}{3}\left(\sinh 2a + \cosh 2a\right)\cos 2b, \tag{4.112}$$

as a solution of Laplace's equation gives, (4.106) gives

$$x = -e^a \cos b + \frac{1}{3}e^{3a} \cos^3 b + e^{3a} \sin^2 b \cos b,$$
$$y = e^a \sin b + \frac{1}{3}e^{3a} \sin^3 b + e^{3a} \cos^2 b \sin b, \tag{4.113}$$
$$u = e^{2a}\left(\cos^2 b - \sin^2 b\right).$$

Identifying that

$$p = -e^a \cos b, \qquad q = -e^a \sin b, \tag{4.114}$$

gives Enneper's minimal surface [6]

$$x = p - \frac{1}{3}p^3 + pq^2,$$
$$y = -q + \frac{1}{3}q^3 - p^2 q, \tag{4.115}$$
$$u = p^2 - q^2.$$

4.3.2.3 Henneberg's Surface

By choosing

$$W = -\frac{2}{3}\cosh 2a \cos 2b, \tag{4.116}$$

as a solution of Laplace's equation gives, (4.106) gives

$$x = \frac{2}{3}\sinh 3a \cos 3b - 2\sinh a \cos b,$$
$$y = \frac{2}{3}\sinh 3a \sin 3b + 2\sinh a \sin b, \tag{4.117}$$
$$u = 2\cosh 2a \cos 2b,$$

known as Henneberg's minimal surface [6].

4.3.2.4 Catalan's Surface

By choosing

$$W = 4 \sinh a \sin b - 2a \cosh a \sin b - 2b \sinh a \cos b, \qquad (4.118)$$

as a solution of Laplace's equation, (4.106) gives obtain

$$x = 2b - \cosh 2a \sin 2b,$$
$$y = 1 - \cosh 2a \cos 2b, \qquad (4.119)$$
$$u = 4 \sinh a \sin b,$$

known as Catalan's minimal surface [6].

4.3.2.5 Bour's Surface

By choosing

$$W = -\frac{1}{2} (\cosh 3a + \sinh 3a) \cos 3b, \qquad (4.120)$$

as a solution of Laplace's equation gives, from (4.106)

$$x = e^{2a} \cos 2b - \frac{e^{4a}}{2} \cos 4b,$$
$$y = -e^{2a} \sin 2b - \frac{e^{4a}}{2} \sin 4b, \qquad (4.121)$$
$$u = \frac{4}{3} e^{3a} \cos 3b.$$

Identifying that $r = e^{2a}$ and $b = \theta/2$ gives rise to

$$x = r \cos \theta - \frac{r^2}{2} \cos 2\theta,$$
$$y = -r \sin \theta - \frac{r^2}{2} \sin 2\theta, \qquad (4.122)$$
$$u = \frac{4}{3} r^{4/3} \cos \frac{3}{2}\theta,$$

the minimal surface known as Bour's surface [6].

Up to this point, we have been able to check whether a transformation is in fact a contact transformation. One just needs to check the contact conditions (4.74). However, trying to create one from scratch can be at times a difficult task. Here we present a fairly straightforward method to generate contact transformations.

4.4   Generating Contact Transformations

The contact conditions introduced earlier this chapter ((4.74) with λ eliminated) are:

$$U_x - PX_x - QY_x = -p\,(U_u - PX_u - QY_u),\qquad\qquad (4.123\text{a})$$

$$U_y - PX_y - QY_y = -q\,(U_u - PX_u - QY_u),\qquad\qquad (4.123\text{b})$$

$$U_p - PX_p - QY_p = 0,\qquad\qquad (4.123\text{c})$$

$$U_q - PX_q - QY_q = 0.\qquad\qquad (4.123\text{d})$$

Verifying that a transformation is indeed a contact transformation is to show that the contact conditions (4.123) are satisfied. This is a fairly straight forward exercise provided that we are given X, Y, U, P and Q. For example, does

$$X = x + \frac{u}{p},\quad Y = y + \frac{u}{q},\quad U = \frac{u^3}{pq},\quad P = \frac{u^2}{q},\quad Q = \frac{u^2}{p},\qquad (4.124)$$

form a contact transformation? A simple substitution of (4.124) into (4.123) show that they are indeed satisfied. But suppose we are given limited information. Could we still verify that we have a contact transformation. For example, does

$$X = u,\quad Y = q,\quad U = x,\qquad\qquad (4.125)$$

form a contact transformation? One could calculate U_X and U_Y directly giving

$$P = \frac{u_{yy}}{u_x u_{yy} - u_y u_{xy}},\quad Q = \frac{u_y}{u_x u_{yy} - u_y u_{xy}},\qquad\qquad (4.126)$$

showing that (4.125) is not a contact transformation. We could have deduced the same conclusion directly from (4.123) as (4.123d) gives $Q = 0$ which is inadmissible. Suppose we try and generalize our choice. For example, does

$$X = A(u),\quad Y = B(q),\quad U = C(x),\qquad\qquad (4.127)$$

form a contract transformation for suitably chosen functions A, B and C, or maybe

$$X = A(x, u),\quad Y = B(x, q),\quad U = C(x, u)?\qquad\qquad (4.128)$$

One can verify the answer is no in both cases. Of course, we could try and generalize our choices further but the analysis would become increasingly more difficult. Here we propose a fairly simple way of generating contact transformations.

Assuming that

$$\frac{\partial(P, Q)}{\partial(p, q)} \neq 0,\qquad\qquad (4.129)$$

we perform a hodograph transformation on the variables (p, q) and (P, Q), i.e.

$$(x, y, u, p, q) \to (x, y, u, P, Q).$$

In doing so (4.123c) and (4.123d) become

$$U_P - P X_P - Q Y_P = 0,$$
$$U_Q - P X_Q - Q Y_Q = 0,$$

(4.130)

while (4.123a) and (4.123b) remain unchanged (the reader should verify this!). If we introduce a new function $F = F(x, y, u, P, Q)$ such that

$$X = F_P, \quad Y = F_Q,$$

(4.131)

then we can integrate (4.130) giving

$$U = P F_P + Q F_Q - F + U_0(x, y, u)$$

(4.132)

where U_0 is an arbitrary function of integration which we can set to zero without loss of generality. Using (4.131) and (4.132), (4.123a) and (4.123b) gives

$$p = -\frac{F_x}{F_u}, \quad q = -\frac{F_y}{F_u},$$

(4.133)

Thus, given $F(x, y, u, P, Q)$ then

$$X = F_P, \quad Y = F_Q, \quad U = p F_P + Q F_Q - F, \quad p = -\frac{F_x}{F_u}, \quad q = -\frac{F_y}{F_u}, \quad (4.134)$$

constitutes a contact transformation provided we are able to solve X, Y, U, P and Q in terms of x, y, u, p and q, and vice-versa. The following examples illustrate.

Example 4.11 If

$$F = x P + y Q - u,$$

(4.135)

then

$$F_x = P, \quad F_y = Q, \quad F_u = -1, \quad F_P = x, \quad F_Q = y,$$

(4.136)

and from (4.134)

$$X = x, \quad Y = y, \quad U = u, \quad p = P, \quad q = Q,$$

(4.137)

so in this case, we get the identity.

Example 4.12 If

$$F = x P + u Q - y,$$

(4.138)

then

$$F_x = P, \quad F_y = -1, \quad F_u = Q, \quad F_P = x, \quad F_Q = u, \tag{4.139}$$

and from (4.134)

$$X = x, \quad Y = u, \quad U = y, \quad p = -\frac{P}{Q}, \quad q = \frac{1}{Q}, \tag{4.140}$$

from which we see

$$P = -\frac{p}{q}, \quad Q = \frac{1}{q} \tag{4.141}$$

a hodograph transformation introduced earlier in this chapter.

Example 4.13 If

$$F = xP^2 + yQ^2 - u, \tag{4.142}$$

then

$$F_x = P^2, \quad F_y = Q^2, \quad F_u = -1, \quad F_P = 2xP, \quad F_Q = 2yQ, \tag{4.143}$$

and from (4.134)

$$X = 2xP, \quad Y = 2yQ, \quad U = xP^2 + yQ^2 + u, \quad p = P^2, \quad q = Q^2, \tag{4.144}$$

from which we obtain the contact transformation

$$X = 2x\sqrt{p}, \quad Y = 2y\sqrt{q}, \quad U = xp + yq + u, \quad P = \sqrt{p}, \quad Q = \sqrt{q}. \tag{4.145}$$

Example 4.14 Determine whether an F exists such that (4.134) derives the following contact transformation

$$X = y + 2\frac{P}{q}, \quad Y = u, \quad U = e^{-x}\frac{P^2}{q^2}. \tag{4.146}$$

It's a fairly simple matter to derive

$$P = e^{-x}\frac{P}{q}, \quad Q = -e^{-x}\frac{P}{q^2}, \tag{4.147}$$

so in fact we do have a contact transformation. We now need to show whether F exists or not, and if so, to find it. From (4.123) we have

$$X = F_P = y + 2\frac{P}{q},$$

$$Y = F_Q = u, \tag{4.148}$$

$$U = PF_P + QF_Q - F = e^{-x}\frac{P^2}{q^2},$$

but since

$$p = -\frac{F_x}{F_u}, \quad q = -\frac{F_y}{F_u}, \tag{4.149}$$

we have

$$F_P = y + 2\frac{F_x}{F_y},$$ (4.150a)

$$F_Q = u,$$ (4.150b)

$$PF_P + QF_Q - F = e^{-x}\frac{F_x^2}{F_y^2}$$ (4.150c)

We integrate (4.150b) giving

$$F = uQ + G(x, y, u, P)$$ (4.151)

where G is an arbitrary function of its arguments. Substituting (4.151) into the remaining 2 equations in (4.150) gives

$$G_P = y + 2\frac{G_x}{G_y},$$ (4.152a)

$$PG_P - G = e^{-x}\frac{G_x^2}{G_y^2}.$$ (4.152b)

Eliminating G_x/G_y from (4.152) gives

$$PG_P - G = \frac{1}{4}e^{-x}(G_P - y)^2,$$ (4.153)

and differentiating (4.153) with respect to P gives

$$PG_{PP} = \frac{1}{2}e^{-x}(G_P - y)G_{PP}$$ (4.154)

from which we deduce

$$G_P = 2e^x P + y, \quad G_{PP} = 0.$$ (4.155)

We only consider the first case and leave it as an exercise to the reader to show the second case is inadmissible. If

$$G_P = 2e^x P + y,$$ (4.156)

then from (4.153) we find

$$G = e^x P^2 + yP$$ (4.157)

and one can verify that indeed that (4.152a) and (4.152b) are satisfied. Thus,

$$F = uQ + e^x P^2 + yP.$$ (4.158)

To verify, if F is given in (4.158) then from (4.134) we have

$$X = F_P = 2e^x P + y,$$
$$Y = F_Q = u,$$
$$U = PF_P + QF_Q - F = e^2 P^2,$$
$$p = -\frac{F_x}{F_u} = -\frac{e^x P^2}{Q},$$
$$q = -\frac{F_y}{F_u} = -\frac{P}{Q}.$$

(4.159)

We solve the latter two for P and Q giving

$$P = \frac{e^{-x} p}{q}, \quad Q = -\frac{e^{-x} p}{q^2},$$

(4.160)

which is exactly (4.147).

Example 4.15 Determine whether an F exists such that (4.134) derives the following contact transformation

$$X = x + u, \quad Y = y + u,$$

(4.161a)

$$P = A\left(x + u, \frac{q+1}{p}\right), \quad Q = B\left(y + u, \frac{p+1}{q}\right).$$

(4.161b)

for some functions A and B. In this example, we are not given U so finding U_X and U_Y for arbitrary $U(x, y, u, p, q)$ would certainly be complicated. We could consider the contact conditions (4.123) with an unknown U but this would be seeking compatibility and unfortunately the form of A and B in (4.161b) are not given so that again would be complicated. Instead we will try and determined whether the method outlined in this section will work. Following (4.134), we have

$$X = F_P = x + u,$$

(4.162a)

$$Y = F_Q = y + u.$$

(4.162b)

As (4.161) would need to be a legit contact transformation then we require that $A_2 \neq 0, B_2 \neq 0$, where differentiation refers to the second arguments, and thus, we can solve (4.161b) giving

$$\frac{q+1}{p} = \tilde{A}(x + u, P), \quad \frac{p+1}{q} = \tilde{B}(y + u, Q)$$

(4.163)

for some functions $\tilde{A}$ and $\tilde{B}$ noting that $\tilde{A}_P \neq 0$ and $\tilde{B}_Q \neq 0$. (At this point we drop the barred variable notation). From (4.134) with $p = -F_x/F_u$ and $q = -F_y/F_u$, (4.163) becomes

$$F_y - F_u = A(x + u, P)F_x, \tag{4.164a}$$
$$F_x - F_u = B(y + u, Q)F_y. \tag{4.164b}$$

Requiring that (4.162) and (4.164) be compatible gives

$$\begin{aligned} A_P F_x + A + 1 &= 0, \\ B_Q F_y + B + 1 &= 0, \end{aligned} \tag{4.165}$$

which gives

$$F_x = -\frac{A+1}{A_P}, \quad F_y = -\frac{B+1}{B_Q}. \tag{4.166}$$

Eliminating F_u in (4.164) and using (4.166) gives

$$\frac{A_P}{(A+1)^2} = \frac{B_Q}{(B+1)^2} \tag{4.167}$$

and since A is a function of $x + u$ and P only and B is a function of $y + u$ and Q only, we conclude that each term in (4.167) is constant and so

$$\frac{A_P}{(A+1)^2} = -k, \quad \frac{B_Q}{(B+1)^2} = -k \tag{4.168}$$

where k is an arbitrary constant which integrate giving

$$A = -1 + \frac{k}{P + a'(x+u)}, \quad B = -1 + \frac{k}{Q + b'(y+u)} \tag{4.169}$$

where a and b are arbitrary functions of their arguments. From (4.162), (4.164), and (4.166) we have

$$\begin{aligned} F_x &= P + a'(x+u), \\ F_y &= Q + b'(y+u), \\ F_u &= P + Q + a'(x+u) + b'(y+u) - k, \\ F_P &= x + u, \\ F_Q &= y + u. \end{aligned} \tag{4.170}$$

We solve this system giving

$$F = (x+u)P + (y+u)Q + a(x+u) + b(y+u) - ku + F_0, \tag{4.171}$$

where F_0 is an arbitrary constant and through (4.134) we find the contact transformation

$$\begin{aligned} X &= x + u, \quad Y = y + u, \quad U = ku - a(x+u) - b(y+u), \\ P &= \frac{kp}{p+q+1} - a'(x+u), \quad Q = \frac{kq}{p+q+1} - b'(y+u). \end{aligned} \tag{4.172}$$

Of course, setting $a = 0, b = 0, k = 1$ certainly would simplify things.

Example 4.16 Determine whether an F exists such that (4.134) derives the following contact transformation

$$P = p - \frac{1}{q}, \quad Q = p + \frac{1}{q}, \tag{4.173a}$$

$$X = A\left(p - \frac{1}{q}, x\left(p - \frac{1}{q}\right) - u\right), \quad Y = B\left(p + \frac{1}{q}, x\left(p + \frac{1}{q}\right) - u\right). \tag{4.173b}$$

for some functions A and B. In this example, we are not given U so finding U_X and U_Y for arbitrary $U(x, y, u, p, q)$ would certainly be complicated. We could consider the contact conditions (4.123) with an unknown U but this would be seeking compatibility and unfortunately the form of A and B in (4.173) are not given (again) so this would be complicated. Instead we will try and determined whether the method outlined in this section will work. In the previous problem, we were given X and Y and P and Q were unknown. Here, P and Q are given and X and Y are unknown. From (4.173a) and (4.134), (4.173b) becomes

$$X = F_P = A(P, xP - u), \tag{4.174a}$$

$$Y = F_Q = B(Q, xQ - u), \tag{4.174b}$$

noting that we have used (4.173a) in (4.173b). Since

$$p = -\frac{F_x}{F_u}, \quad q = -\frac{F_y}{F_u} \tag{4.175}$$

then from (4.173a)

$$2F_x + (P + Q)F_u = 0, \tag{4.176a}$$

$$(Q - P)F_y + 2F_u = 0. \tag{4.176b}$$

Compatibility of (4.174) and (4.176) gives

$$F_x = \frac{1}{2}(P^2 - Q^2)A_{\bar{P}}, \quad F_x = -\frac{1}{2}(P^2 - Q^2)B_{\bar{Q}},$$

$$F_y = -2A_{\bar{P}}, \quad F_y = 2B_{\bar{Q}}, \tag{4.177}$$

$$F_u = (Q - P)A_{\bar{P}}, \quad F_u = (P - Q)B_{\bar{Q}},$$

where $\bar{P} = xP - u$ and $\bar{Q} = xQ - u$. Further compatibility of (4.174) and (4.177) gives

$$A_{P\bar{P}} = A_{\bar{P}\bar{P}} = B_{Q\bar{Q}} = B_{\bar{Q}\bar{Q}} = 0, \tag{4.178}$$

from which we solve giving

$$A = c_1 (xP - u) + A_0'(P), \quad B = -c_1(xQ - u) + B_0'(Q), \tag{4.179}$$

where c_1 is an arbitrary constant and A_0 and B_0 are arbitrary functions of their arguments with primes denoting differentiation. With this, we are able to solve (4.174) and (4.177) explicitly for F giving

$$F = \frac{c_1}{2} \left(P^2 - Q^2\right) x + c_1(Q - P)u - 2c_1 y + A_0(P) + B_0(Q). \tag{4.180}$$

If we suppress the functions A_0 and B_0 and set $c_1 = 1$, through (4.134) we find the contact transformation

$$X = x \left(p - \frac{1}{q}\right) - u, \quad Y = -x \left(p + \frac{1}{q}\right) + u, \quad U = 2y - 2\frac{xp}{q}. \tag{4.181}$$

Up to this point, we have considered contact transformations involving single PDEs and a single dependent variable. Varley [10] was able to extend Legendre transformations to a class of first order systems of PDEs. This we consider next.

4.5   Parametric Legendre Transformations

Here we consider the first order system of PDEs

$$F(u_x, u_y, v_x, v_y) = 0, \quad G(u_x, u_y, v_x, v_y) = 0, \tag{4.182}$$

where F and G are smooth functions of their arguments. In 1962 Varley [10] was able to show that it is possible to linearize (4.182). Suppose that (4.182) can be written parametrically as

$$\begin{aligned} u_x &= p(a, b), & u_y &= q(a, b), \\ v_x &= r(a, b), & v_y &= s(a, b), \end{aligned} \tag{4.183}$$

for some parameters (a, b). If we introduce the Legendre transformations

$$U = xp + yq - u, \quad V = xr + ys - v, \tag{4.184}$$

then

$$dU = xdp + ydq, \quad dV = xdr + yds, \tag{4.185}$$

since

$$du = pdx + qdy, \quad dv = rdx + sdy. \tag{4.186}$$

Furthermore, expressing the differentials in (4.185) in terms of the parameter variables (a, b), we obtain

$$U_a da + U_b db = x \left(p_a da + p_b db \right) + y \left(q_a da + q_b db \right)$$
$$V_a da + V_b db = x \left(r_a da + r_b db \right) + y \left(s_a da + s_b db \right) \tag{4.187}$$

from which we obtain comparing coefficients da and db

$$x p_a + y q_a = U_a, \qquad x p_b + y q_b = U_b,$$
$$x r_a + y s_a = V_a, \qquad x r_b + y s_b = V_b. \tag{4.188}$$

Using (4.183), we solve any two of (4.188) for x and y and the remaining 2 become linear PDEs for U and V. The following examples illustrate.

Example 4.17 In the study of two dimensional dilatant fluids, Varley [11] obtained the equations

$$u_x + v_y = 0,$$
$$\frac{1}{2} \left(u_x^2 + v_y^2 \right) + \frac{1}{4} \left(u_y + v_x \right)^2 = 1. \tag{4.189}$$

We introduce the parameterizations

$$u_x = p = -\sin 2a, \qquad u_y = q = b + \cos 2a,$$
$$v_x = r = -b + \cos 2a, \qquad v_y = s = \sin 2a. \tag{4.190}$$

Using (4.190), Eq. (4.188) becomes

$$-2x \cos 2a - 2y \sin 2a = U_a, \qquad y = U_b,$$
$$-2x \sin 2a + 2y \cos 2a = V_a, \qquad -x = V_b. \tag{4.191}$$

from which we see that
$$x = -V_b, \quad y = U_b, \tag{4.192}$$

and the two remaining equations in (4.191) become

$$2 \cos 2a \, V_b - 2 \sin 2a \, U_b = U_a,$$
$$2 \sin 2a \, V_b + 2 \cos 2a \, U_b = V_a, \tag{4.193}$$

a pair of linear PDEs for the unknowns U and V. Once solved for U and V, (4.192) gives x and y while (4.184) with (4.190) gives

$$u = V_a + b U_b - U, \quad v = -U_a + b V_b - V. \tag{4.194}$$

It is interesting to note that via the substitution

$$U = \cos a \, \bar{U} - \sin a \, \bar{V}, \quad V = \sin a \, \bar{U} + \cos a \, \bar{V} \tag{4.195}$$

(4.193) become the constant coefficients PDEs

$$\bar{U}_a - 2\bar{V}_b - \bar{V} = 0,$$
$$\bar{V}_a - 2\bar{U}_b + \bar{U} = 0.$$
(4.196)

Example 4.17 The following appears in modeling isotropic compressible hyperelastic materials [12]

$$u_x u_y + v_x v_y = 0,$$
$$u_x^2 - u_y^2 + v_x^2 - v_y^2 = 1.$$
(4.197)

We introduce the parameterizations

$$u_x = p = \cosh a \cos b, \qquad u_y = q = -\sinh a \sin b,$$
$$v_x = r = \cosh a \sin b, \qquad v_y = s = \sinh a \cos b.$$
(4.198)

Using (4.198), Eq. (4.188) becomes

$$x \sinh a \cos b - y \cosh a \sin b = U_a, \qquad -x \cosh a \sin b - y \sinh a \cos b = U_b,$$
$$x \sinh a \sin b + y \cosh a \cos b = V_a, \qquad x \cosh a \cos b - y \sinh a \sin b = V_b,$$
(4.199)

from which we see that

$$x = \frac{\cos b U_a + \sin b V_a}{\sinh a}, \qquad y = \frac{-\sin b U_a + \cos b V_a}{\cosh a}$$
(4.200)

and the two remaining equations in (4.199) become

$$\sin 2b U_a + \sinh 2a U_b + (\cosh 2a - \cos 2b) V_a = 0,$$
$$(\cosh 2a + \cos 2b) U_a + \sin 2b V_a - \sinh 2a V_b = 0$$
(4.201)

a pair of linear PDEs for the unknowns U and V. Once solved, (4.200) give x and y while (4.184) with (4.198) gives

$$u = V_b - U, \qquad v = -U_b - V.$$
(4.202)

It is interesting to note that solving (4.201) for V_a and V_b gives

$$V_a = \frac{\sin 2b\, U_a + \sinh 2a\, U_b}{\cos 2b - \cosh 2a}, \qquad V_b = \frac{\sin 2b\, U_b - \sinh 2a\, U_a}{\cos 2b - \cosh 2a}$$
(4.203)

and eliminating V gives that U satisfies Laplace's equation

$$U_{aa} + U_{bb} = 0.$$
(4.204)

Here, we consider two solutions of (4.204) showing how to can obtain exact solutions to (4.197). In particular, we consider

$$U = c \cosh a \sin b, \quad U = c \sinh a \cos b, \tag{4.205}$$

where c is an arbitrary constant.

Solution 1 If

$$U = c \cosh a \sin b, \tag{4.206}$$

then we solve (4.203) for V giving

$$V = -c \cosh a \cos b + \frac{c}{2} \ln \left| \frac{\cosh a + \cos b}{\cosh a - \cos b} \right|, \tag{4.207}$$

where the constant of integration is suppressed. With these ((4.206) and (4.207)), x, y, u and v in (4.200) and (4.202) become

$$x = -c \frac{\sin b \cos b}{\cosh^2 a - \cos^2 b}, \quad y = -c \frac{\sinh a \cosh a}{\cosh^2 a - \cos^2 b}, \tag{4.208}$$
$$u = -c \frac{\cosh a \sin b}{\cosh^2 a - \cos^2 b}, \quad v = -\frac{c}{2} \ln \left| \frac{\cosh a + \cos b}{\cosh a - \cos b} \right|.$$

As the original PDE (4.197) admits a scaling symmetry we have the freedom to choose c as we like without loss of generality so we choose $c = -1$ giving

$$x = \frac{\sin b \cos b}{\cosh^2 a - \cos^2 b}, \quad y = \frac{\sinh a \cosh a}{\cosh^2 a - \cos^2 b}, \tag{4.209}$$
$$u = \frac{\cosh a \sin b}{\cosh^2 a - \cos^2 b}, \quad v = \frac{1}{2} \ln \left| \frac{\cosh a + \cos b}{\cosh a - \cos b} \right|.$$

Eliminating a and b in (4.209) gives rise to

$$u^4 + (y^2 - x^2 - 1)u^2 - x^2 y^2 = 0, \tag{4.210}$$
$$y^2 \tanh^4 v + (x^2 - y^2 + 1) \tanh^2 v - x^2 = 0.$$

We could go further and solve (4.210) explicitly for u and v in terms of x and y and this is left as an exercise for the reader.

Solution 2 If

$$U = c \sinh a \cos b, \tag{4.211}$$

then we solve (4.203) for V giving

$$V = c \left(\sinh a \sin b + \tan^{-1} \left(\frac{\sin b}{\sinh a} \right) \right). \tag{4.212}$$

With these ((4.211) and (4.212)), x, y, u and v in (4.200) and (4.202) become

$$x = c\frac{\sinh a \cosh a}{\cosh^2 a - \cos^2 b}, \qquad y = -c\frac{\sin b \cos b}{\cosh^2 a - \cos^2 b},$$

$$u = c\frac{\sinh a \cos b}{\cosh^2 a - \cos^2 b}, \qquad v = -c\tan^{-1}\left(\frac{\sin b}{\sinh a}\right).$$

(4.213)

As the original PDE (4.197) admits several scaling symmetries it suffices to consider

$$x = \frac{\sinh a \cosh a}{\cosh^2 a - \cos^2 b}, \qquad y = \frac{\sin b \cos b}{\cosh^2 a - \cos^2 b},$$

$$u = \frac{\sinh a \cos b}{\cosh^2 a - \cos^2 b}, \qquad v = \tan^{-1}\left(\frac{\sin b}{\sinh a}\right).$$

(4.214)

Eliminating a and b in (4.214) gives rise to

$$u^4 + (y^2 - x^2 + 1)u^2 - x^2 y^2 = 0,$$
$$x^2 \tan^4 v + (x^2 - y^2 - 1)\tan^2 v - y^2 = 0.$$

(4.215)

We could go further and solve (4.215) explicitly for u and v in terms of x and y and again, this is left as an exercise for the reader.

In the two previous examples, we provided the parametrization. The inquisitive reader may ask how does one come up with such a parametrization and is it unique? We will try and answer these questions via the following example.

Example 4.18 Consider the system of PDEs

$$u_x + v_y = 0, \tag{4.216a}$$
$$u_x v_y - u_y v_x = 1, \tag{4.216b}$$

that arises in modeling incompressible hyperelastic materials in nonlinear elasticity [2]. We wish to linearize this system via parametric Legendre transformations. Here we consider two parameterizations. For example, if we choose $u_x = a$, then from (4.216a) $v_y = -a$. With these, (4.216b) becomes

$$-a^2 - u_y v_x = 1 \tag{4.217}$$

If we choose $u_y = b$ then $v_x = -(1 + a^2)/b$ and we have our first paramaterization

$$u_x = a, \quad u_y = b, \quad v_x = -\frac{1 + a^2}{b}, \quad v_y = -a. \tag{4.218}$$

For our second parameterization, substituting (4.216a) into (4.216b) gives

$$-u_x^2 - u_y v_x = 1 \tag{4.219}$$

As we have the identity $\cosh^2 a - \sinh^2 a = 1$ we choose

$$u_x = \sinh a, \quad u_y v_x = -\cosh^2 a \tag{4.220}$$

from which we choose (not unique)

$$u_y = b \cosh a, \quad v_x = -\frac{\cosh a}{b}. \tag{4.221}$$

Thus, we have as our second parameterization

$$u_x = \sinh a, \quad u_y = b \cosh a, \quad v_x = -\frac{\cosh a}{b}, \quad v_y = -\sinh a. \tag{4.222}$$

We follow through with the first set of parameterizations and leave it to the exercises for the second set.

With the parameterization (4.218), Eq. (4.188) becomes

$$
\begin{aligned}
x &= U_a, & y &= U_b, \\
-\frac{2ax}{b} - y &= V_a, & \frac{1+a^2}{b^2} x &= V_b,
\end{aligned} \tag{4.223}
$$

which gives

$$x = U_a, \quad y = U_b \tag{4.224}$$

and the remaining two equations in (4.223) becoming

$$
\begin{aligned}
2aU_a + bU_b + bV_a &= 0 \\
(a^2 + 1)U_a - b^2 V_b &= 0
\end{aligned} \tag{4.225}
$$

From (4.184) (using (4.225)) , we have

$$u = -aU_a - bV_a - U, \quad v = -aU_b - bV_b - V. \tag{4.226}$$

It is interesting to note that under the point transformation

$$
\begin{aligned}
a &= \tan \bar{a}, & b &= \frac{e^{\bar{b}}}{\cos \bar{a}} \\
U &= \frac{e^{\bar{b}} \bar{U}}{\cos \bar{a}}, & V &= -\tan \bar{a} \bar{U} + \bar{V}
\end{aligned} \tag{4.227}
$$

the system (4.225) become the Cauchy-Riemann equations

$$\bar{U}_{\bar{a}} - \bar{V}_{\bar{b}} = 0, \quad \bar{U}_{\bar{b}} + \bar{V}_{\bar{a}} = 0. \tag{4.228}$$

4.6   Exercises

1. Show the following are contact transformations.

$$\text{(i)}\quad x = X + Y + U_X, \quad y = X + 3U_X, \quad u = 2XU_X - 2U + X^2 + Y^2,$$

$$\text{(ii)}\quad x = e^{-Y}, \quad y = U, \quad u = X^2 e^{-Y},$$

$$\text{(iii)}\quad x = Y + \frac{2U_X}{U_Y}, \quad y = U, \quad u = e^{-X} \frac{U_X^2}{U_Y^2},$$

$$\text{(iv)}\quad x = U_X, \quad y = \frac{1}{U_Y}, \quad u = \frac{U - XU_X - YU_Y}{U_Y}.$$

2. Given X, Y, P and Q, show a U exists such that the contact conditions are satisfied; further, find U

$$\text{(i)}\quad X = u, \quad Y = y - \frac{xp}{q}, \quad P = -\frac{x}{q}, \quad Q = x,$$

$$\text{(ii)}\quad X = x - 2p, \quad Y = y, \quad P = p, \quad Q = q,$$

$$\text{(iii)}\quad X = q, \quad Y = y - \frac{u}{q}, \quad P = \frac{u}{pq}, \quad Q = -\frac{q}{p},$$

$$\text{(iv)}\quad X = x + \frac{1}{p}, \quad Y = y + \frac{1}{q}, \quad P = A(p,q), \quad Q = B(p,q).$$

In part (iv), you will need to determine the forms A and B.

3. Show that the following is a contact transformation.

$$t = 2T, \quad x = \frac{2U - (T + X)u_X}{2(T + X)}, \quad u = \frac{U - (T + X)U_X}{2(T + X)^2}. \tag{4.229}$$

Further show that the Hunter-Saxon equation [3]

$$u_{tx} + uu_{xx} + \frac{1}{2}u_x^2 = 0, \tag{4.230}$$

is transformed to [4]

$$U_{TX} = \frac{2(U_T + U_X)}{T + X} - \frac{4U}{(T + X)^2}. \tag{4.231}$$

4. The boundary Layer equations from fluid mechanics are

$$u_x + v_y = 0, \tag{4.232a}$$

$$uu_x + vu_y = u_{yy}. \tag{4.232b}$$

4 (i) Show that (4.232b) can be written as

$$\left(u^2\right)_x + \left(uv - u_y\right)_y = 0. \tag{4.233}$$

4 (ii) Show that with the introduction of the stream functions $\psi = \psi(x, y)$ and $\phi = \phi(x, y)$ such that

$$u = \psi_y, \quad v = -\psi_x, \quad u^2 = \phi_y, \quad uv - u_y = -\phi_x, \tag{4.234}$$

(4.232a) and (4.233) are automatically satisfied. Furthermore, from (4.234), ψ and ϕ satisfy

$$\psi_y^2 = \phi_y, \tag{4.235a}$$

$$\psi_{yy} + \psi_x \psi_y = \phi_x. \tag{4.235b}$$

4 (iii) Show under the Hodograph transformation,

$$x = X, \quad y = \Psi, \quad \psi = Y, \quad \phi = \Phi, \tag{4.236}$$

system (4.235) becomes

$$\Phi_Y = \frac{1}{\Psi_Y}, \quad \Phi_X = -\frac{\Psi_{YY}}{\Psi_Y^3}, \tag{4.237}$$

which, on eliminating Φ, gives

$$\Psi_Y \Psi_{XY} - \Psi_X \Psi_{YY} - \frac{2\Psi_{YY}^2}{\Psi_Y^3} + \frac{\Psi_{YY}}{\Psi_Y^2} = 0, \tag{4.238}$$

or

$$\left(\frac{1}{\Psi_Y}\right)_X = -\left(\frac{\Psi_{YY}}{\Psi_Y^3}\right)_Y. \tag{4.239}$$

Under the Hodograph transformation (4.236), u becomes

$$u = \frac{1}{\Psi_Y}, \tag{4.240}$$

giving (4.239) as

$$u_X = (uu_Y)_Y \tag{4.241}$$

a nonlinear diffusion equation! This has been noted in Ames [5].

4 (iv) Show under the Hodograph transformation,

$$x = X, \quad y = \Phi, \quad \psi = \Psi, \quad \phi = Y, \tag{4.242}$$

system (4.235) becomes

$$\frac{\Psi_Y^2}{\Phi_Y^2} - \frac{1}{\Phi_Y} = 0, \quad \frac{\Psi_{YY}}{\Phi_Y^2} - \frac{\Psi_Y \Phi_{YY}}{\Phi_Y^3} + \frac{\Psi_X \Psi_Y}{\Phi_Y} - \frac{\Psi_Y^2 \Phi_X}{\Phi_Y^2} + \frac{\Phi_X}{\Phi_Y} = 0, \tag{4.243}$$

which on simplifying gives

$$\Psi_Y^2 - \Phi_Y = 0, \quad \Psi_X - \frac{\Psi_{YY}}{\Psi_Y^3} = 0. \tag{4.244}$$

Differentiating the second in (4.244) gives

$$(\Psi_Y)_X = \left(\frac{\Psi_{YY}}{\Psi_Y^3}\right)_Y. \tag{4.245}$$

Show under the Hodograph transformation (4.242), u becomes

$$u = \frac{1}{\Psi_Y}, \tag{4.246}$$

giving (4.245) as

$$(u^*)_X = \left(\frac{u^*_Y}{u^{*3}}\right)_Y, \tag{4.247}$$

where $u^* = u^{-1}$, another nonlinear diffusion equation! Rogers and Shadwick have shown that different nonlinear diffusion equations can be linked via reciprocal transformations [9].

5. The Harry-Dym equation is given by

$$u_t = u^3 u_{xxx}.$$

Show that by introducing the change of variable $u = v^{-1/2}$, this equation becomes (to within scaling of t)

$$v_t = \left(\frac{v_x}{v^{3/2}}\right)_{xx},$$

which, under the substitution $v = w_x$, becomes

$$w_t = \left(\frac{w_{xx}}{w_x^{3/2}}\right)_x. \tag{4.248}$$

Show under a Hodograph transformation, (4.248) remains unchanged.

6. The equations for irrotational steady 2-D flows in fluid mechanics are given by

$$(\rho u)_x + (\rho v)_y = 0, \tag{4.249a}$$

$$u u_x + v u_y = -\frac{p_x}{\rho}, \tag{4.249b}$$

$$u v_x + v v_y = -\frac{p_y}{\rho}, \tag{4.249c}$$

$$\nabla \times \mathbf{v} = 0. \tag{4.249d}$$

(i) Show that for 2 - D flows, (4.249d) is $u_y - v_x = 0$, and is identically satisfied by introducing a potential $u = \phi_x$ and $v = \phi_y$.

(ii) Assuming that $p = c^2\rho$ where c is a constant speed of sound, show that the system reduces to the single equation

$$\left(\phi_x^2 - c^2\right)\phi_{xx} + 2\phi_x\phi_y\phi_{xy} + \left(\phi_y^2 - c^2\right)\phi_{yy} = 0. \tag{4.250}$$

(iii) Show that under a Legendre transformation this equation linearizes.

7. The governing equations for highly frictional granular materials is given by

$$\frac{\partial\sigma_{xx}}{\partial x} + \frac{\partial\sigma_{xy}}{\partial y} = 0, \tag{4.251a}$$

$$\frac{\partial\sigma_{xy}}{\partial x} + \frac{\partial\sigma_{yy}}{\partial y} = \rho g, \tag{4.251b}$$

$$\sigma_{xx}\sigma_{yy} - \sigma_{xy}^2 = 0, \tag{4.251c}$$

where σ_{xx}, σ_{yy} and σ_{xy} are normal and shear stresses, respectively, ρ the material density, and g, the gravity. In (4.251), (4.251a) and (4.251b) are the equilibrium equations, whereas (4.251c) is the constitutive relation for the material [7].

(i) Show that by introducing a potential u, such that

$$\sigma_{xx} = u_y, \quad \sigma_{xy} = -u_x, \quad \text{then } \sigma_{yy} = \frac{u_x^2}{u_y}, \tag{4.252}$$

Equation (4.251) reduces to

$$u_y^2 u_{xx} - 2u_x u_y u_{xy} + u_x^2 u_{yy} + \rho g u_y^2 = 0. \tag{4.253}$$

(ii) Show that under a Hodograph transformation, this linearizes.

8. The shallow water wave equations are given by

$$u_t + uu_x + g\eta_x = 0, \tag{4.254a}$$
$$\eta_t + [u(\eta - h(x))]_x = 0, \tag{4.254b}$$

where $u = u(x, t)$ the velocity, $\eta = \eta(x, t)$ the free surface, $h = h(x)$ the bottom surface, and g is gravity [8].

(i) Introducing a potential function ϕ where

$$\eta = \phi_x + h(x), \quad u = -\frac{\phi_t}{\phi_x}, \tag{4.255}$$

show the shallow water equations become

$$\phi_x^2 \phi_{tt} - 2\phi_t \phi_x \phi_{tx} + \left(\phi_t^2 - g\phi_x^3\right)\phi_{xx} - gh'(x)u_x^3 = 0. \tag{4.256}$$

(ii) Show that under a modified Hodograph transformation, this equation for $h(x) = -\alpha x$ can be transformed to

$$U_X^3 U_{TT} - gU_{XX} = 0. \tag{4.257}$$

Further show that under a Legendre transformation, this equation can be linearized. Compose the two transformations.

9. Derive contact transformations from the given F

$$(i) \quad F = xP + yQ - u + \frac{P}{Q}$$

$$(ii) \quad F = uP + (x + u)Q - xy$$

$$(iii) \quad F = \frac{xP}{u} + \frac{yQ}{u} - u$$

10. Can the following contact transformations be derived from

$$X = F_P, \quad Y = F_Q, \quad U = pF_P + QF_Q - F, \quad p = -\frac{F_x}{F_u}, \quad q = -\frac{F_y}{F_u},$$

(i) $X = xp^2, \quad Y = yq^2, \quad U = 2xp + 2yq - u, \quad P = 1/p, \quad Q = 1/q$

(ii) $X = q, \quad Y = u/q - y, \quad U = x, \quad P = u/pq, \quad Q = q/p$

(iii) $X = x - p, \quad Y = y, \quad U = 2u + \dfrac{x^2 - 2xp - p^2}{2}, P = x + p, \quad Q = 2q$

(iv) $X = x - 2y + p, \quad Y = x - y + p, \quad U = xp - u + \dfrac{1}{2}x^2 - \dfrac{1}{2}y^2,$

$\quad P = y - x + q, \quad Q = 2x - y - q$

(v) $X = x - 2y + p, \quad Y = y - 2x + q, \quad U = pq - u - (x - y)(p - q) - \dfrac{3}{2}(x - y)^2,$

$\quad P = y - x + q, \quad Q = x - y + p$

11. Can the following contact transformations be derived from

$$X = F_P, \quad Y = F_Q, \quad U = pF_P + QF_Q - F, \quad p = -\frac{F_x}{F_u}, \quad q = -\frac{F_y}{F_u},$$

(i) $\quad X = xp^2, \quad Y = yq^2, \quad U = 2xp + 2yq - u, \quad P = 1/p, \quad Q = 1/q$

(ii) $\quad X = q, \quad Y = u/q - y, \quad U = x, \quad P = u/pq, \quad Q = q/p$

(iii) $\quad X = x - p, \quad Y = y, \quad U = 2u + \dfrac{x^2 - 2xp - p^2}{2}, P = x + p, \quad Q = 2q$

(iv) $\quad X = x - 2y + p, \quad Y = x - y + p, \quad U = xp - u + \dfrac{1}{2}x^2 - \dfrac{1}{2}y^2,$

$\quad\quad P = y - x + q, \quad Q = 2x - y - q$

(v) $\quad X = x - 2y + p, \quad Y = y - 2x + q, \quad U = pq - u - (x - y)(p - q) - \dfrac{3}{2}(x - y)^2,$

$\quad\quad P = y - x + q, \quad Q = x - y + p$

12. In this section we considered

$$u_x + v_y = 0,$$
$$u_x v_y - u_y v_x = 1.$$

A parameterization was given in (4.222), namely

$$u_x = \sinh a, \quad u_y = b \cosh a, \quad v_x = -\frac{\cosh a}{b}, \quad v_y = -\sinh a.$$

Show

$$x = \frac{V_a}{\cosh b}, \quad y = \frac{a^2 U_a}{\cosh b}, \quad u = \frac{V_b}{a} - U, \quad v = a U_b - V,$$

where U and V satisfy

$$a \sinh b U_a + \cosh b U_b - \cosh b V_a = 0,$$
$$a^2 \cosh b U_a - a \sinh b V_a + \cosh b V_b = 0.$$

13. Linearize

$$v_x = \frac{u_y}{u_x^2 + u_y^2}, \quad v_y = \frac{u_x}{u_x^2 + u_y^2},$$

References

1. S. Lie, *Göttinger Nachrichten* (1872), p. 480
2. D.J. Arrigo, J.M. Hill, Transformations and equation reductions in finite elasticity. I. Plane strain deformations. Math. Mech. Solids **1**(2), 155–175 (1996)
3. J.K. Hunter, R. Saxton, Dynamics of director fields. SIAM J. Appl. Math. **51**(6), 1498–1521 (1991)
4. O.L. Morozov, Contact Equivalence of the Generalized Hunter - Saxton Equation and the Euler - Poisson Equation, arXiv:math-ph/0406016v2
5. W.F. Ames *Nonlinear Partial Differential Equations in Engineering*, vol. I (1968)

6. E.W. Weisstein, *Minimal Surfaces* From MathWorld–A Wolfram Web Resource. http://mathworld.wolfram.com
7. D.J. Arrigo, L. Le, J.W. Torrence, Exact solutions for a class of ratholes in highly frictional granular materials. Dyn. Cont. Dis. Imp. Syst.: Appl. Alg. **19**, 497–509 (2010)
8. J.J. Stoker, *Water Waves - The Mathematical Theory with Applications, Wiley-Interscience* (Interscience, New York, 1958)
9. C. Rogers, W.F. Shadwick, *Bäcklünd Transformations and Their Applications* (Academic, 1982)
10. E. Varley, A class of non-linear partial differential equations. Commun. Pure Appl. Math. **15**, 91–94 (1962)
11. E. Varley, *Flows in Dilatant Fluids*, Brown University Ph.D. thesis (1961)
12. D.J. Arrigo, T.C. Chism, Symmetry classification of plane deformations of hyperelastic compressible solids. Math. Mech. Solids **0**(0), (2022). https://doi.org/10.1177/10812865221117475

First Integrals

5

In an introductory course in PDEs, the wave equation

$$u_{tt} = c^2 u_{xx}, \tag{5.1}$$

where $c > 0$ is a constant wave speed, is introduced. The general solution of (5.1) is

$$u = F(x - ct) + G(x + ct), \tag{5.2}$$

where F and G are arbitrary functions of their arguments. In this solution, there are two waves; one is traveling right ($F(x - ct)$) and one is traveling left ($G(x + ct)$). Each of these solutions can be obtained from the following first order PDEs

$$u_t + cu_x = 0, \tag{5.3a}$$
$$u_t - cu_x = 0. \tag{5.3b}$$

PDE (5.3a) gives rise to $u = F(x - ct)$, whereas PDE (5.3b) gives rise to $u = G(x + ct)$. We question whether it is possible to show this directly—whether the solutions of (5.3a) and/or (5.3b) will give rise to solutions of (5.1). For example, differentiating (5.3a) with respect to t and x gives

$$u_{tt} + cu_{tx} = 0,$$
$$u_{tx} + cu_{xx} = 0. \tag{5.4}$$

It is a simple matter to show that upon elimination of u_{tx} in (5.4) we obtain (5.1). The same follows from (5.3b). In this example, we considered the linear wave equation, where the solution of a lower order PDE gave rise to solutions of a higher order PDE. Does this idea apply for nonlinear PDEs? The following example illustrates the concept.

Consider the following pair of PDEs:

$$u_x u_y = 1, \tag{5.5a}$$

$$u_{xx} - u_y^{-4} u_{yy} = 0. \tag{5.5b}$$

We ask, will solutions of (5.5a) give rise to solutions of (5.5b)? For example, one exact solution of (5.5a) is $u = 2\sqrt{xy}$. Calculating all necessary derivatives gives

$$u_x = \frac{\sqrt{y}}{\sqrt{x}}, \quad u_y = \frac{\sqrt{x}}{\sqrt{y}}, \quad u_{xx} = -\frac{\sqrt{y}}{2x\sqrt{x}}, \quad u_{yy} = -\frac{\sqrt{x}}{2y\sqrt{y}}, \tag{5.6}$$

and substituting into (5.5b) shows it is identically satisfied. The reader can also verify that $u = x + y$ also satisfies both PDEs. We naturally ask, will all of the solutions of (5.5a) give rise to solutions (5.5b)? It's a simple matter to calculate the derivatives of (5.5a)

$$u_x = \frac{1}{u_y} \quad u_{xx} = -\frac{u_{xy}}{u_y^2} \quad u_{xy} = -\frac{u_{yy}}{u_y^2} \tag{5.7}$$

and eliminating u_{xy} from (5.7) gives (5.5b). If a PDE of lower order

$$F(x, y, u, u_x, u_y) = 0 \tag{5.8}$$

exists such that differential consequences of (5.8) gives rise to the original (higher order) PDE, then the lower order PDE is called a **first integral**. We see that (5.5a) is a first integral of (5.5b). Are there others and, if so, how do we find them? If we assume a general form as in (5.8), then differential consequences give

$$F_x + F_u u_x + F_p u_{xx} + F_q u_{xy} = 0, \tag{5.9a}$$

$$F_y + F_u u_y + F_p u_{xy} + F_q u_{yy} = 0, \tag{5.9b}$$

where $p = u_x$ and $q = u_y$. Eliminating u_{xy} from (5.9) gives

$$F_p^2 u_{xx} - F_q^2 u_{yy} + F_q (F_x + p F_u) - F_p (F_y + q F_u). \tag{5.10}$$

In order to obtain our targeted PDE (5.5b), we need to solve the following system of PDE for F

$$F_p^2 - q^4 F_q^2 = 0, \tag{5.11a}$$

$$F_q (F_x + p F_u) - F_p (F_y + q F_u) = 0. \tag{5.11b}$$

Since (5.11a) factors $(F_p - q^2 F_q)(F_p + q^2 F_q) = 0$, there are two cases to consider. We consider the first case here and leave it to the reader to consider the second case. Solving

$$F_p - q^2 F_q = 0 \tag{5.12}$$

gives

$$F = G\left(x, y, u, p - \frac{1}{q}\right),\tag{5.13}$$

where G is an arbitrary function of its arguments. Substituting (5.13) into (5.11b) and letting $\lambda = p - 1/q$ gives

$$(G_x + \lambda G_u)\, q^2 - G_y = 0.\tag{5.14}$$

Since $G = G(x, y, u, \lambda)$ and is now independent of q, then from (5.14) we have both

$$\begin{aligned} G_x + \lambda G_u &= 0,\\ G_y &= 0, \end{aligned}\tag{5.15}$$

which are easily solved, giving $G = H(\lambda, x\lambda - u)$ where H is an arbitrary function of its arguments, and the final form of F is

$$F = H_1\left(p - \frac{1}{q}, x\left(p - \frac{1}{q}\right) - u\right).\tag{5.16}$$

The second case is solved the same way and gives rise to

$$F = H_2\left(p + \frac{1}{q}, x\left(p + \frac{1}{q}\right) - u\right).\tag{5.17}$$

As we can see from (5.16) and (5.17), there is actually an enormous class of first integrals to (5.8). As we will see later, PDEs admitting general classes of first integrals sometimes have important properties.

5.1 Quasilinear Second Order Equations

In this section we consider general second order quasilinear PDEs in two independent variables

$$Ar + Bs + Ct = E,\tag{5.18}$$

where $r = u_{xx}, s = u_{xy}, t = u_{yy}$ and A, B, C and E are functions of x, y, u, u_x and u_y. We assume a first integral of the form (5.8). Differentiating with respect to x and y (see (5.9)) and isolating u_{xx} and u_{yy} gives

$$r = -\frac{F_x + pF_u + sF_q}{F_p}, \quad t = -\frac{F_y + qF_u + sF_p}{F_q},\tag{5.19}$$

where we are assuming that $F_p F_q \neq 0$. Substituting (5.19) into (5.18) and simplifying gives

$$\left(AF_q^2 - BF_pF_q + CF_p^2\right)s + AF_q\left(F_x + pF_u\right) + CF_p\left(F_y + qF_u\right) + EF_pF_q = 0\tag{5.20}$$

and, since F is independent of s, this leads to

$$AF_q^2 - BF_p F_q + CF_p^2 = 0, \tag{5.21a}$$

$$AF_q (F_x + pF_u) + CF_p (F_y + qF_u) + EF_p F_q = 0. \tag{5.21b}$$

Note that (5.21a) is quadratic in F_q/F_p and can be solved depending on whether $B^2 - 4AC > 0$, $B^2 - 4AC = 0$ or $B^2 - 4AC < 0$. We have assumed that $F_p F_q \neq 0$. Cases where either $F_p = 0$ or $F_q = 0$ would be considered special cases and would need to be considered separately.

Example 5.1 Obtain a first integral to

$$u_y^2 u_{xx} - 2u_x u_y u_{xy} + u_x^2 u_{yy} = u_y^3. \tag{5.22}$$

Identifying $A = q^2$, $B = -2pq$, $C = p^2$, and $E = q^3$, from the first integral equations (5.21) gives

$$q^2 F_q^2 + 2pq F_p F_q + p^2 F_p^2 = 0, \tag{5.23a}$$

$$q^2 F_q (F_x + pF_u) + p^2 F_p (F_y + qF_u) + q^3 F_p F_q = 0. \tag{5.23b}$$

We find (5.23a) a perfect square, which gives

$$pF_p + qF_q = 0, \tag{5.24}$$

which simplifies (5.23b) to

$$qF_q (qF_x - pF_y + q^2 F_p) = 0, \tag{5.25}$$

or

$$qF_x - pF_y + q^2 F_p = 0. \tag{5.26}$$

Note that we have excluded the case where $F_q = 0$ as we have assumed that $F_p F_q \neq 0$. The solution of (5.24) is

$$F = G\left(x, y, u, \frac{p}{q}\right), \tag{5.27}$$

and substituting into (5.26) and letting $p = \lambda q$ gives

$$G_x - \lambda G_y + G_\lambda = 0, \tag{5.28}$$

whose solution is

$$G = H\left(u, x - \lambda, y - \frac{1}{2}x^2 + \lambda x\right). \tag{5.29}$$

Thus, (5.22) admits first integral of the form

$$H\left(u, x - \frac{u_x}{u_y}, y - \frac{1}{2}x^2 + \frac{xu_x}{u_y}\right) = 0,$$ (5.30)

where H is an arbitrary function of its arguments. For example, if we choose

$$u_x - xu_y = 0,$$ (5.31)

then we obtain the exact solution $u = f(x^2 + 2y)$ of the original PDE. Other choices of F in (5.30) could lead to exact solutions. For example, choosing

$$2xu_x + (2y - x^2)u_y = 0,$$ (5.32)

would lead to the exact solution $u = F\left(\frac{y}{x} + \frac{x}{2}\right)$. As a final possibility, we choose

$$\frac{u_x}{u_y} - x = f(u).$$ (5.33)

This is very much like (5.31) involving an arbitrary function $f(u)$ and leads to the exact solution

$$y + \frac{1}{2}x^2 = -f(u)x + g(u),$$ (5.34)

involving now two arbitrary functions f and g.

Example 5.2 Obtain a first integral to

$$\left(u_y^2 + u_y\right)u_{xx} - \left(2u_xu_y + u_x\right)u_{xy} + u_x^2u_{yy} = 0.$$ (5.35)

Here $A = q^2 + q$, $B = -2pq - p$, $C = p^2$, and $E = 0$, giving from the first integral equations (5.21)

$$\left(q^2 + q\right)F_q^2 + (2pq + p)F_pF_q + p^2F_p^2 = 0,$$ (5.36a)
$$\left(q^2 + q\right)F_q\left(F_x + pF_u\right) + p^2F_p\left(F_y + qF_u\right) = 0.$$ (5.36b)

From (5.36a), we find

$$\left(pF_p + qF_q\right)\left(pF_p + (1 + q)F_q\right) = 0,$$ (5.37)

showing that there are two cases to consider.

Case 1: If
$$pF_p + qF_q = 0,$$ (5.38)

then from (5.36b) we obtain the second equation

$$(q + 1)F_x - pF_y + pF_u = 0,$$ (5.39)

and the solution of the over-determined system (5.38) and (5.39) is

$$F = G_1\left(y + u, \frac{p}{q}\right).$$
(5.40)

Case 2: If

$$pF_p + (q + 1)F_q = 0,$$
(5.41)

then from (5.36b) we obtain the second equation

$$qF_x - pF_y = 0,$$
(5.42)

and the solution of the over-determined system (5.41) and (5.42) is

$$F = G_2\left(u, \frac{p}{q + 1}\right).$$
(5.43)

From either (5.40) or (5.43) we can find the general solution of (5.35). For example, if we choose from (5.40)

$$\frac{p}{q} = f(u + y),$$
(5.44)

where f is arbitrary, if we let $f = -1/\phi'$, the solution of (5.44) is

$$x = \phi(y + u) + \psi(u),$$
(5.45)

where ϕ and ψ are arbitrary functions of their arguments.

Example 5.3 Consider

$$\left(1 + u_y^2\right)u_{xx} - 2u_x u_y u_{xy} + u_x^2 u_{yy} = 0.$$
(5.46)

Here $A = 1 + q^2$, $B = -2pq$, $C = p^2$, $E = 0$ and our first integral equations are

$$(1 + q^2)F_q^2 + 2pq\,F_p F_q + p^2 F_p^2 = 0,$$
(5.47a)

$$(1 + q^2)F_q(F_x + pF_u) + p^2 F_p(F_y + qF_u) = 0.$$
(5.47b)

We solve (5.47a), giving

$$F = G\left(x, y, u, \frac{q \pm i}{p}\right),$$
(5.48)

where i is the usual $i = \sqrt{-1}$. Requiring that (5.48) satisfies (5.49) gives the final form

$$F = H\left(u \mp iy, \frac{q \pm i}{p}\right).$$
(5.49)

One will notice that the first integrals presented in (5.49) are complex. However, it is possible to construct real solutions to (5.46). For example, if we choose

$$\frac{q+i}{p} = u - iy, \tag{5.50a}$$

$$\frac{q-i}{p} = u + iy, \tag{5.50b}$$

then we can split (5.50) into real and imaginary parts giving

$$q = up, \tag{5.51a}$$
$$yp = -1, \tag{5.51b}$$

leading to the common solution

$$u = \frac{c - x}{y}. \tag{5.52}$$

In this example, c is an arbitrary constant and one can verify that (5.52) satisfies (5.46).

If we choose

$$\frac{q+i}{p} = \frac{1}{u - iy}, \tag{5.53a}$$

$$\frac{q-i}{p} = \frac{1}{u + iy}. \tag{5.53b}$$

then we can split (5.53) into real and imaginary parts, giving

$$uq + y = p, \tag{5.54a}$$
$$u - yq = -1, \tag{5.54b}$$

leading to the solution

$$u = \tan(x + c)y. \tag{5.55}$$

Finally, choosing

$$\frac{q+i}{p} = (u - iy)^2, \tag{5.56a}$$

$$\frac{q-i}{p} = (u + i)^2, \tag{5.56b}$$

then splitting (5.56) into real and imaginary parts gives

$$q = (u^2 - y^2)p, \tag{5.57a}$$
$$2yup = -1, \tag{5.57b}$$

leading to the solution

$$u^2 = \frac{y^2}{2} - \frac{x}{y} + c. \tag{5.58}$$

The reader can verify that (5.55) and (5.58) are both solutions of (5.46).

Example 5.4 Obtain a first integral to

$$u_{xx} + 2uu_{xy} + u^2 u_{yy} = 0. \tag{5.59}$$

Identifying that $A = 1$, $B = 2u$, $C = u^2$, and $E = 0$, gives from the first integral equations (5.21)

$$F_q^2 - 2u F_p F_q + u^2 F_p^2 = 0, \tag{5.60a}$$

$$F_q (F_x + p F_u) + u^2 F_p (F_y + q F_u) = 0. \tag{5.60b}$$

From (5.60a) we find

$$F_q - u F_p = 0, \tag{5.61}$$

which simplifies (5.60b) to

$$u F_p (F_x + u F_y + (p + uq) F_u) = 0 \tag{5.62}$$

or

$$F_x + u F_y + (p + uq) F_u = 0. \tag{5.63}$$

Note that we have excluded the case where $F_p = 0$ as we have assumed that $F_p F_q \neq 0$. The solution of (5.61) is

$$F = G (x, y, u, p + uq). \tag{5.64}$$

Substituting this into (5.63) and letting $\lambda = p + uq$ gives

$$G_x + u G_y + \lambda G_u + q \lambda G_\lambda = 0. \tag{5.65}$$

Since G is independent of q, this gives

$$\lambda G_\lambda = 0. \tag{5.66}$$

Normally we would say that $G_\lambda = 0$ which would then give that F is independent of p and q, i.e., no first integral. But we also have the choice that $\lambda = 0$, and one can show that, in fact, $\lambda = p + uq = 0$ is a first integral.

5.2   Monge-Ampere Equation

In this section we consider Monge-Ampere equations. These are second order PDEs in two independent variables in the form

$$Ar + Bs + Ct + D\left(rt - s^2\right) = E, \tag{5.67}$$

where in addition to the forms of A, B, C, and E in the previous section, we also assume that $D \neq 0$ and is a function of x, y, u, u_x, and u_y. We again assume a first integral of the form (5.8) and on isolating the derivatives r and t and substituting into (5.67) and isolating coefficients with respect to s gives.

$$AF_q^2 - BF_p F_q + CF_p^2 = D\left(F_x F_p + F_y F_q + pF_u F_p + qF_u F_q\right), \tag{5.68a}$$

$$AF_q\left(F_x + pF_u\right) + CF_p\left(F_y + qF_u\right) + EF_p F_q = D\left(F_x F_y + pF_y F_u + qF_x F_u + pqF_u^2\right). \tag{5.68b}$$

In an attempt to solve (5.68), we assume there exists a $\lambda(x, y, u, p, q)$, such that

$$\lambda\left(AF_q^2 - BF_p F_q + CF_p^2 - D\left(F_x F_p + F_y F_q + pF_u F_p + qF_u F_q\right)\right)$$
$$+AF_q\left(F_x + pF_u\right) + CF_p\left(F_y + qF_u\right) + EF_p F_q - D\left(F_x F_y + pF_y F_u + qF_x F_u + pqF_u^2\right) \tag{5.69}$$
$$= \left(A_1 F_x + B_1 F_u + C_1 F_p + D_1 F_q\right)\left(A_2 F_y + B_2 F_u + C_2 F_p + D_2 F_q\right).$$

Expanding (5.69) and equating coefficients of the product of F_x, F_y, F_u, F_p, F_q gives

$$A_1 A_2 = -D, \quad A_2 B_1 = -pD, \quad A_1 B_2 = -qD,$$
$$B_1 B_2 = -pqD, \quad A_1 C_2 = -\lambda D, \quad D_1 A_2 = -\lambda D,$$
$$A_2 C_1 = C, \quad A_1 D_2 = A, \quad C_1 C_2 = \lambda C, \quad D_1 D_2 = \lambda A, \tag{5.70}$$
$$B_1 D_2 + D_1 B_2 = pA - \lambda qD, \quad B_1 C_2 + C_1 B_2 = qC - p\lambda D,$$
$$C_1 D_2 + D_1 C_2 = -\lambda B + E,$$

from which we find the following

$$B_1 = pA_1, \quad C_1 = -\frac{A_1 C}{D}, \quad D_1 = \lambda A_1,$$

$$A_2 = -\frac{D}{A_1}, \quad B_2 = -\frac{qD}{A_1}, \quad C_2 = -\frac{\lambda D}{A_1}, \quad D_2 = \frac{A}{A_1}, \tag{5.71}$$

where λ satisfies

$$D^2\lambda^2 - BD\lambda + AC + DE = 0. \tag{5.72}$$

Solving (5.72) for λ gives rise to the following possibilities:

(i)   two real distinct roots,
(ii) two real repeated roots,
(iii) two complex roots.

In the case of two distinct roots, say λ_1 and λ_2, (5.69) can be factored as

$$\left(DF_x + pDF_u - CF_p + \lambda_1 DF_q\right)\left(DF_y + qDF_u + \lambda_1 DF_p - AF_q\right) = 0, \qquad (5.73a)$$
$$\left(DF_x + pDF_u - CF_p + \lambda_2 DF_q\right)\left(DF_y + qDF_u + \lambda_2 DF_p - AF_q\right) = 0. \qquad (5.73b)$$

There are 4 systems to be considered, however, two of these give $\lambda_1 = \lambda_2$; as we have assumed that $\lambda_1 \neq \lambda_2$, these are inadmissible. Thus, we shall only consider the systems

$$D\left(F_x + pF_u\right) - CF_p + \lambda_1 DF_q = 0, \qquad (5.74a)$$
$$D\left(F_y + qF_u\right) + \lambda_2 DF_p - AF_q = 0, \qquad (5.74b)$$

and

$$D\left(F_x + pF_u\right) - CF_p + \lambda_2 DF_q = 0, \qquad (5.75a)$$
$$D\left(F_y + qF_u\right) + \lambda_1 DF_p - AF_q = 0. \qquad (5.75b)$$

In the case of a repeated root, we only have one set of equations for F to consider, as (5.74) and (5.75) coalesce into one. The following examples illustrate.

Example 5.5 Obtain first integral to

$$r + 3s + t + rt - s^2 = 1. \qquad (5.76)$$

Identifying that $A = 1$, $B = 2$, $C = 1$, $D = 1$, and $E = 1$, then (5.72) gives

$$\lambda^2 - 3\lambda + 2 = 0, \qquad (5.77)$$

leading to $\lambda = 1, 2$. Thus, we have two cases to consider.

Case 1: From (5.74) we have

$$F_x + pF_u - F_p + F_q = 0, \qquad (5.78a)$$
$$F_y + qF_u + 2F_p - F_q = 0. \qquad (5.78b)$$

Using the method of characteristic, we solve the first, giving

$$F = G\left(x + p, y, u + \frac{1}{2}p^2, x - q\right). \qquad (5.79)$$

Substituting into the second gives

$$G_y + 2G_\alpha + (2\alpha - \gamma - x)G_\beta + G_\gamma = 0, \qquad (5.80)$$

where $\alpha = x + p$, $\beta = u + \frac{1}{2}p^2$, and $\gamma = x - q$. With these variables, G is independent of x; this leads to

$$G_\beta = 0, \quad G_y + 2G_\alpha + G_\gamma = 0, \qquad (5.81)$$

from which we obtain the final form of first integral

$$F = H_1(x - 2y + p, y - x + q). \qquad (5.82)$$

Case 2: From (5.75) we have

$$F_x + pF_u - F_p + 2F_q = 0, \qquad (5.83a)$$
$$F_y + qF_u + F_p - F_q = 0. \qquad (5.83b)$$

Using the method of characteristic, we solve the first, giving

$$F = G\left(x + p, y, u + \frac{1}{2}p^2, 2x - q\right). \qquad (5.84)$$

Substituting into the second gives

$$G_y + G_\alpha + (\alpha - \gamma + x)G_\beta + G_\gamma = 0, \qquad (5.85)$$

where $\alpha = x + p$, $\beta = u + \frac{1}{2}p^2$, and $\gamma = 2x - q$. With these variables, G is independent of x; this leads to

$$G_\beta = 0, \quad G_y + G_\alpha + G_\gamma = 0, \qquad (5.86)$$

from which we obtain the final form of first integral

$$F = H_2(x - y + p, y - 2x + q). \qquad (5.87)$$

Any choice of H in (5.82) or (5.87) will lead to an exact solution of the original PDE. For example, from (5.87), choosing

$$a(x - y + p) + b(y - 2x + q) = 0, \qquad (5.88)$$

where a and b are arbitrary constants and f is an arbitrary function, gives

$$u = \frac{b^2 - ab - a^2}{2a^2}x^2 + \frac{2a - b}{a}xy + f(bx - ay), \qquad (5.89)$$

an exact solution to (5.76).

Example 5.6 Obtain first integral to

$$xqr - (x + y)s + ypt + xy\left(rt - s^2\right) = 1 - pq. \tag{5.90}$$

Identifying that $A = xq$, $B = -(x + y)$, $C = yp$, $D = xy$, and $E = 1 - pq$, then from (5.72) gives

$$x^2y^2\lambda^2 + xy(x + y)\lambda + xypq + xy(1 - pq) = 0, \tag{5.91}$$

leading to $\lambda = -\dfrac{1}{x}, -\dfrac{1}{y}$. Thus, we have two cases to consider.

Case 1: From (5.74) we have

$$xF_x + xpF_u - pF_p - F_q = 0, \tag{5.92a}$$
$$yF_y + yqF_u - F_p - qF_q = 0. \tag{5.92b}$$

Using the method of characteristic, we solve the first, giving

$$F = G\left(xp, y, u - xp\ln x, \ln x + q\right). \tag{5.93}$$

Substituting into the second gives

$$G_y + y(\gamma - \ln x)G_\beta - xG_\alpha + x\ln xG_\beta + (\ln x - \gamma)\, G_\gamma = 0, \tag{5.94}$$

or, re-arranging gives

$$G_y + y\gamma G_\beta - \gamma G_\gamma - xG_\alpha + \ln x\left(G_\gamma - yG_\beta\right) + x\ln xG_\beta = 0, \tag{5.95}$$

where $\alpha = x + p$, $\beta = u + \frac{1}{2}p^2$, and $\gamma = x - q$. With these variables, G is independent of x and thus leads to

$$G_y = 0, \quad G_\alpha = 0, \quad G_\beta = 0, \quad G_\gamma = 0, \tag{5.96}$$

gives that F is constant. So in this case, there is no first integral.

Case 2: From (5.75) we have

$$xyF_x + xypF_u - ypF_p - xF_q = 0, \tag{5.97a}$$
$$xyF_y + xyqF_u - yF_p - xqF_q = 0. \tag{5.97b}$$

Using the method of characteristic, we solve the first, giving

$$F = G\left(xp, y, u + xp\ln x, x + yq\right). \tag{5.98}$$

Substituting (5.98) into (5.97b) and re-arranging gives

$$\left(G_y - yG_\alpha - \gamma G_\beta\right)x + G_\beta x^2 - xy\ln xG_\beta = 0, \tag{5.99}$$

where $\alpha = xp$, $\beta = u + xp \ln x$, and $\gamma = x + yq$. With these variables, G is independent of x; this leads to

$$G_\beta = 0, \quad G_y - G_\alpha = 0, \tag{5.100}$$

from which we obtain $G = H(\gamma, y + \alpha)$ and the final form of first integral

$$F = H(xp + y, yq + x), \tag{5.101}$$

where H is an arbitrary function. Any choice of H in (5.101) will lead to an exact solution of the original PDE. For example, choosing

$$xp + yq + x + y = 0 \tag{5.102}$$

leads to the exact solution

$$u = -(x + y) + f(x/y), \tag{5.103}$$

where f is arbitrary as an exact solution to (5.90).

Example 5.7 Obtain first integral to

$$yqr + 2xys + y^2 t + y^2(rt - s^2) = x^2 - yq. \tag{5.104}$$

Identifying that $A = yq$, $B = 2xy$, $C = y^2$, $D = y^2$, and $E = x^2 - yq$, from (5.72) we get

$$y^4\lambda^2 - 2xy^3\lambda + x^2y^2 = 0, \tag{5.105}$$

leading to the repeated root $\lambda = x/y$. Thus, we have only a single case to consider. From (5.74) we have

$$y F_x + yp F_u - y F_p + x F_q = 0, \tag{5.106a}$$

$$y F_y + yq F_u + x F_p - q F_q = 0. \tag{5.106b}$$

Using the method of characteristic, we solve the first, giving

$$F = G\left(x + p, y, 2u + p^2, 2yq - x^2\right). \tag{5.107}$$

Substituting into the second gives

$$y G_y + \gamma G_\beta + (G_\alpha + 2\alpha G_\beta) x - G_\beta x^2 = 0, \tag{5.108}$$

where $\alpha = x + p$, $\beta = 2u + p^2$, and $\gamma = 2yq - x^2$. With these variables, G is independent of x; this leads to

$$G_\beta = 0, \quad G_\alpha + 2\alpha G_\beta = 0, \quad y G_y + \gamma G_\beta = 0, \tag{5.109}$$

from which we obtain $G = H(\gamma)$, leading to the final form of first integral

$$F = H(2yq - x^2).\tag{5.110}$$

For example, if we choose

$$2yu_y - x^2 = 0\tag{5.111}$$

we integrate, giving

$$u = \frac{1}{2}x^2 \ln|y| + C(x),\tag{5.112}$$

where $C(x)$ is an arbitrary function and substitution into (5.104) shows it is exactly satisfied.

Example 5.8 Consider

$$u_{xx} + u_{yy} + 2\left(u_{xx}u_{yy} - u_{xy}^2\right) = 0.\tag{5.113}$$

Here

$$A = 1, \quad B = 0, \quad C = 1, \quad D = 2, \quad E = 0,\tag{5.114}$$

and from (5.72) we have

$$4\lambda^2 + 1 = 0,\tag{5.115}$$

giving $\lambda = \pm i/2$. Thus, (5.74) and (5.75) become

$$2F_x + 2pF_u - F_p - iF_q = 0,\tag{5.116a}$$
$$2F_y + 2qF_u + iF_p - F_q = 0,\tag{5.116b}$$

and

$$2F_x + 2pF_u - F_p + iF_q = 0,\tag{5.117a}$$
$$2F_y + 2qF_u - iF_p - F_q = 0.\tag{5.117b}$$

The solutions of each are

$$F_1(x + 2u_x - iy, u_x + iu_y) = 0,$$
$$F_2(x + 2u_x + iy, u_x - iu_y) = 0.\tag{5.118}$$

Again, one will notice that the first integrals in (5.118) are complex. However, it is again possible to construct real solutions to (5.113). For example, if we choose

$$x + 2u_x - iy + u_x + iu_y = 0,\tag{5.119a}$$
$$x + 2u_x + iy + u_x - iu_y = 0,\tag{5.119b}$$

then we can split into real and imaginary parts, giving

$$x + 3u_x = 0, \tag{5.120a}$$

$$y - u_y = 0, \tag{5.120b}$$

which leads to the solution

$$u = -\frac{x^2}{6} + \frac{y^2}{2} + c, \tag{5.121}$$

where c is an arbitrary constant.

If we choose

$$(x + 2u_x - iy)^2 + u_x + iu_y = 0, \tag{5.122a}$$

$$(x + 2u_x + iy)^2 + u_x - iu_y = 0, \tag{5.122b}$$

then we can split into real and imaginary parts, giving

$$4yu_x - u_y + 2xy = 0, \tag{5.123a}$$

$$4u_x^2 + (4x + 1)u_x + x^2 - y^2 = 0, \tag{5.123b}$$

which leads to the common solution

$$u = -\frac{2x^2 + 2y^2 + x}{8} \pm \frac{(1 + 8x + 16y^2)^{3/2}}{96} + c, \tag{5.124}$$

where c is an arbitrary constant. The reader can verify that (5.121) and (5.124) do indeed satisfy (5.113).

5.3   The Martin Equation

The governing equations of an inviscid, one-dimensional nonsteady gas, neglecting heat condition and heat radiation are

$$\rho_t + (\rho u)_x = 0, \tag{5.125a}$$

$$u_t + uu_x = -\frac{P_x}{\rho}, \tag{5.125b}$$

or, in conservative form

$$\rho_t + (\rho u)_x = 0, \tag{5.126a}$$

$$(\rho u) + \left(\rho u^2 + P\right)_x = 0. \tag{5.126b}$$

Normally a third equation is given, conservation of energy or an equation of state but here, we will leave the system underdetermined. If we introduce stream functions ψ and $\bar{\xi}$ such that

$$\rho = \psi_x, \qquad \rho u = -\psi_t, \tag{5.127a}$$

$$\rho u = \bar{\xi}_x, \quad \rho u^2 + P = -\bar{\xi}_t, \tag{5.127b}$$

then (5.126a) and (5.126b) are automatically satisfied. If we introduce a new variable ξ such that

$$\bar{\xi} = \xi - tP, \tag{5.128}$$

then (5.127b) becomes

$$\rho u = \xi_x - tP_x, \quad \rho u^2 = \xi_t + tP_t. \tag{5.129}$$

Following Martin [3], we choose new independent variables (ψ, P), so that (5.127a) and (5.129) become

$$\rho = -\frac{t_P}{J}, \tag{5.130a}$$

$$\rho u = -\frac{x_P}{J}, \tag{5.130b}$$

$$\rho u = \frac{\xi_P t_\psi - \xi_\psi t_P - t t_\psi}{J}, \tag{5.130c}$$

$$\rho u^2 = \frac{\xi_P x_\psi - \xi_\psi x_P - t x_\psi}{J}, \tag{5.130d}$$

where $J = t_\psi x_P - t_P x_\psi$. From which (5.130a) and (5.130b) we deduce

$$u = \frac{x_P}{t_P}. \tag{5.131}$$

Eliminating ρ and u from (5.130c) and (5.130d) using (5.130a) and (5.131) gives

$$x_P = t t_\psi + t_P \xi_\psi - t_\psi \xi_P, \tag{5.132a}$$

$$\frac{x_P^2}{t_P} = t x_\psi + x_P \xi_\psi - x_\psi \xi_P, \tag{5.132b}$$

from which we deduce

$$(t - \xi_P)\left(t_\psi x_P - t_P x_\psi\right) = 0. \tag{5.133}$$

Thus,

$$t = \xi_P, \tag{5.134}$$

as we are assuming that $t_\psi x_P - t_P x_\psi \neq 0$. From (5.132) we obtain

$$x_P = \xi_\psi \xi_{PP}, \tag{5.135}$$

and from (5.134) and (5.135) gives (5.131) as

$$u = \xi_\psi. \tag{5.136}$$

We return to (5.130a) and with (5.134) this becomes

$$x_\psi = \xi_\psi \xi_{P\psi} + \tau, \tag{5.137}$$

where $\tau = \rho^{-1}$. Eliminating x from (5.135) and (5.137) gives the Monge-Ampere equation

$$\xi_{\psi\psi} \xi_{PP} - \xi_{\psi P}^2 = \tau_P. \tag{5.138}$$

With a renaming of variables, Martin [4] asked, which forms of the Monge-Ampere equation

$$u_{xx} u_{yy} - u_{xy}^2 + \lambda^2 = 0, \quad \lambda = X(x)Y(y), \tag{5.139}$$

admits first integrals. He was able to deduce the following forms.

$$\lambda = Y(y), \quad \lambda = \frac{x^{m-1}}{y^{m+1}}, \quad \lambda = e^x e^y \tag{5.140}$$

m and Y arbitrary with special cases

$$\lambda = y^{-2}, \quad \lambda = x^{-1} y^{-1}. \tag{5.141}$$

We now consider this same problem, however, we will remove the assumption that λ is separable. From the section on first integrals, if a first integral $F(x, y, u, p, q)$ exists, then F satisfies

$$F_x + p F_u - \lambda(x, y) F_q = 0, \tag{5.142a}$$
$$F_y + q F_u + \lambda(x, y) F_p = 0. \tag{5.142b}$$

Requiring that these two be compatible gives rise to the additional equation

$$F_u - \frac{1}{2} \frac{\lambda_x}{\lambda} F_p - \frac{1}{2} \frac{\lambda_y}{\lambda} F_q = 0. \tag{5.143}$$

If we further require that (5.142) and (5.143) be compatible, we obtain

$$\left(\frac{\lambda_{xx}}{\lambda_x} - \frac{3}{2} \frac{\lambda_x}{\lambda} \right) F_p + \left(\frac{\lambda_{xy}}{\lambda_x} - \frac{3}{2} \frac{\lambda_y}{\lambda} \right) F_q = 0, \tag{5.144a}$$

$$\left(\frac{\lambda_{xy}}{\lambda_y} - \frac{3}{2} \frac{\lambda_x}{\lambda} \right) F_p + \left(\frac{\lambda_{yy}}{\lambda_y} - \frac{3}{2} \frac{\lambda_y}{\lambda} \right) F_q = 0. \tag{5.144b}$$

Since $F_p F_q \neq 0$, from (5.144) we obtain

$$\left(\frac{\lambda_{xx}}{\lambda_x} - \frac{3}{2} \frac{\lambda_x}{\lambda} \right) \left(\frac{\lambda_{yy}}{\lambda_y} - \frac{3}{2} \frac{\lambda_y}{\lambda} \right) - \left(\frac{\lambda_{xy}}{\lambda_x} - \frac{3}{2} \frac{\lambda_y}{\lambda} \right) \left(\frac{\lambda_{xy}}{\lambda_y} - \frac{3}{2} \frac{\lambda_x}{\lambda} \right) = 0, \tag{5.145}$$

or

$$\left(2\lambda\lambda_{xx} - 3\lambda_x^2\right)\left(2\lambda\lambda_{yy} - 3\lambda_y^2\right) - \left(2\lambda\lambda_{xy} - 3\lambda_x\lambda_y\right)^2 = 0. \tag{5.146}$$

Melshenko [5] also obtains (5.146) and states that this equation admits solutions of the form

$$\lambda = H\left(c_1 x + c_2 y\right) \quad \text{and} \quad \lambda = (y + c_2)^{-2} H\left(\frac{x + c_1}{y + c_2}\right), \tag{5.147}$$

for arbitrary constants c_1, c_2 and arbitrary function H. Remarkably, under the substitution

$$\lambda = \frac{1}{\omega^2} \tag{5.148}$$

Equation (5.146) becomes the homogeneous Monge-Ampere equation

$$\omega_{xx}\omega_{yy} - \omega_{xy}^2 = 0, \tag{5.149}$$

which is know to admit the solutions ([6])

$$\omega = f\left(c_1 x + c_2 y\right) + c_3 x + c_4 y + c_5, \tag{5.150}$$

for arbitrary constants $c_1 - c_5$ and arbitrary function f,

$$\omega = (c_1 x + c_2 y + c_3) f\left(\frac{c_4 x + c_5 y + c_6}{c_1 x + c_2 y + c_3}\right) + c_7 x + c_8 y + c_9, \tag{5.151}$$

with arbitrary constants $c_1 - c_9$ and arbitrary function f, and the parametric solutions

$$\omega = tx + f(t)y + g(t), \quad x + f'(t)y + g'(t) = 0, \tag{5.152}$$

with arbitrary functions f and g noting that (5.147) are special cases of (5.150) and (5.151). One can also show that (5.149) also admits solutions of the form

$$\omega = (c_1 x + c_2 y + c_3) f\left(\frac{c_4 x + c_5 y + c_6}{c_7 x + c_8 y + c_9}\right) + c_{10} x + c_{11} y + c_{12}, \tag{5.153}$$

with arbitrary constants $c_1 - c_{12}$ and arbitrary function f, but with the constraint that $c_1 c_5 c_9 + c_2 c_6 c_7 + c_3 c_4 c_8 - c_1 c_6 c_8 - c_2 c_4 c_9 - c_3 c_5 c_7 = 0$.

5.4   First Integrals and Linearization

As we've seen in the preceding section, some PDEs admit very general classes of first integrals. A natural question is: can they be used to simplify the form of a PDE?

5.4.1   Hyperbolic MA Equations

Lie ([1] and [2]) considered PDEs of the form

$$Ar + Bs + Ct + D(rt - s^2) = E, \tag{5.154}$$

that admitted two general first integrals, say

$$F_1(\alpha_1, \beta_1) = 0, \quad F_2(\alpha_2, \beta_2) = 0. \tag{5.155}$$

In (5.155), it is assumed that F_1 and F_2 are arbitrary functions of α_i and β_i that are functions of x, y, u, p, and q. Lie was able to show that when these first integrals exist, it is possible to transform (5.154) to $U_{XY} = 0$.

We define

$$X = \alpha_1, \quad Y = \alpha_2, \quad P = \beta_1, \quad Q = \beta_2 \tag{5.156}$$

and we'll assume that U exists, such that the contact conditions are satisfied. We consider

$$U_{XY} = 0.$$

This can be rewritten as

$$\frac{\partial P}{\partial Y} = \frac{\partial(X, P)}{\partial(X, Y)} = \frac{\partial(X, P)}{\partial(x, y)} \bigg/ \frac{\partial(X, Y)}{\partial(x, y)} = 0,$$

giving

$$\frac{\partial(X, P)}{\partial(x, y)} = 0 \tag{5.157}$$

or

$$X_x P_y - X_y P_x = 0. \tag{5.158}$$

We now bring in (5.156) and suppress subscripts. Thus (5.158) becomes

$$\left(\alpha_x + p\alpha_u + r\alpha_p + s\alpha_q\right)\left(\beta_y + q\beta_u + s\beta_p + t\beta_q\right)$$
$$-\left(\alpha_y + q\alpha_u + s\alpha_p + t\alpha_q\right)\left(\beta_x + p\beta_u + r\beta_p + s\beta_q\right) = 0. \tag{5.159}$$

Expanding (5.159) gives

$$\begin{aligned}
&\left[\alpha_p\left(\beta_y + q\beta_u\right) - \beta_p\left(\alpha_y + q\alpha_u\right)\right]r \\
&+ \left[\beta_q\left(\alpha_x + p\alpha_u\right) + \alpha_q\left(\beta_y + q\beta_u\right) - \beta_q\left(\alpha_y + q\alpha_u\right) - \alpha_p\left(\beta_x + p\beta_u\right)\right]s \\
&+ \left[\beta_q\left(\alpha_x + p\alpha_u\right) - \alpha_q\left(\beta_x + p\beta_u\right)\right]t \\
&+ \left(\alpha_p\beta_q - \alpha_q\beta_p\right)\left(rt - s^2\right) \\
&= \left(\alpha_y + q\alpha_u\right)\left(\beta_x + p\beta_u\right) - \left(\alpha_x + p\alpha_u\right)\left(\beta_y + q\beta_u\right).
\end{aligned} \tag{5.160}$$

As α and β are first integrals, they satisfy the following

$$D\left(F_x + pF_u\right) - CF_p + \lambda_1 DF_q = 0, \tag{5.161a}$$
$$D\left(F_y + qF_u\right) + \lambda_2 DF_p - AF_q = 0, \tag{5.161b}$$

where

$$D^2\lambda^2 - BD\lambda + AC + DE = 0. \tag{5.162}$$

As we are assuming that $D \neq 0$, then using (5.161), we eliminate the terms

$$\alpha_x + p\alpha_u, \quad \alpha_y + q\alpha_u, \quad \beta_x + p\beta_u, \quad \beta_y + q\beta_u \tag{5.163}$$

from (5.160). Simplifying gives

$$\frac{A}{D}\left(\alpha_p\beta_q - \beta_p\alpha_q\right)r + (\lambda_1 + \lambda_2)\left(\alpha_p\beta_q - \beta_p\alpha_q\right)s$$
$$+ \frac{C}{D}\left(\alpha_p\beta_q - \beta_p\alpha_q\right)t + \left(\alpha_p\beta_q - \alpha_q\beta_p\right)\left(rt - s^2\right)$$
$$= \left(\lambda_1\lambda_2 - \frac{AC}{D^2}\right)\left(\alpha_p\beta_q - \beta_p\alpha_q\right).$$

As we have assumed that $\alpha_p\beta_q - \beta_p\alpha_q \neq 0$, this term cancels, giving

$$\frac{A}{D}r + (\lambda_1 + \lambda_2)\,s + \frac{C}{D}t + \left(rt - s^2\right) = \left(\lambda_1\lambda_2 - \frac{AC}{D^2}\right).$$

From (5.162), we deduce that $\lambda_1 + \lambda_2 = B/D$ and $\lambda_1\lambda_2 = (AC + DE)/D^2$, giving

$$\frac{A}{D}r + \frac{B}{D}s + \frac{C}{D}t + \left(rt - s^2\right) = \left(\frac{AC + DE}{D^2} - \frac{AC}{D^2}\right). \tag{5.164}$$

Simplifying gives rise to (5.154).

Recall the contact conditions from the previous chapter. They are:

$$U_p - PX_p - QY_p = 0,$$
$$U_q - PX_q - QY_q = 0,$$
$$U_u - PX_u - QY_u = \lambda, \tag{5.165}$$
$$U_x - PX_x - QY_x = -\lambda p,$$
$$U_y - PX_y - QY_y = -\lambda q.$$

These will be needed for the following examples.

Example 5.9 Consider

$$3r + s + t + rt - s^2 + 9 = 0. \tag{5.166}$$

This PDE admits the first integrals

$$F_1(x + 2y + p, 3x - 3y - q) = 0, \quad F_2(x - 3y + p, 2x + 3y + q) = 0. \tag{5.167}$$

Here we try

$$X = x + 2y + p, \quad Y = x - 3y + p, \tag{5.168a}$$
$$P = 3x - 3y - q, \quad Q = 2x + 3y + q. \tag{5.168b}$$

At this point, we need to find U such that the contact conditions (5.165) are satisfied, namely

$$
\begin{aligned}
U_p &= 5x, \\
U_q &= 0, \\
U_u &= \lambda, \\
U_x &= 5x - \lambda p, \\
U_y &= -15y - 5q - \lambda q.
\end{aligned} \tag{5.169}
$$

Cross differentiation of (5.169) shows they are consistent; thus U (and λ) exist. Integrating (5.169) gives

$$U = 5xp + \frac{5}{2}x^2 - \frac{15}{2}y^2 - 5u, \tag{5.170}$$

and calculating U_{XY} using (5.168) and (5.170) gives

$$U_{XY} = \frac{3u_{xx} + u_{xy} + u_{yy} + u_{xx}u_{yy} - u_{xy}^2 + 9}{5\,(u_{xx} + 1)} = 0. \tag{5.171}$$

Example 5.10 Consider

$$u_y^2 u_{xx} - 3u_x u_y u_{xy} + 2u_x^2 u_{yy} = u_x u_y^2. \tag{5.172}$$

This PDE admits the first integrals

$$F_1\left(y + 2\frac{p}{q}, e^{-x}\frac{p}{q}\right) = 0, \quad F_2\left(u, e^{-x}\frac{p}{q^2}\right) = 0. \tag{5.173}$$

Here we try

$$X = y + 2\frac{p}{q}, \quad Y = u, \tag{5.174a}$$
$$P = e^{-x}\frac{p}{q}, \quad Q = e^{-x}\frac{p}{q^2}. \tag{5.174b}$$

At this point, we need to find U such that the contact conditions (5.165) are satisfied, namely

$$U_p = 2e^{-x}\frac{p}{q^2}, \tag{5.175a}$$

$$U_q = -2e^{-x}\frac{p^2}{q^3}, \tag{5.175b}$$

$$U_u = e^{-x}\frac{p}{q^2} + \lambda, \tag{5.175c}$$

$$U_x = -\lambda p, \tag{5.175d}$$

$$U_y = e^{-x}\frac{p}{q} - \lambda q. \tag{5.175e}$$

Cross differentiation of (5.175) shows they are not consistent, so we modify our choice. We now pick

$$X = y + 2\frac{p}{q}, \quad Y = u, \tag{5.176a}$$

$$P = a\,e^{-x}\frac{p}{q}, \quad Q = b\,e^{-x}\frac{p}{q^2}, \tag{5.176b}$$

with hopefully suitably chosen constants a and b. The contact conditions become

$$U_p = 2ae^{-x}\frac{p}{q^2}, \tag{5.177a}$$

$$U_q = -2ae^{-x}\frac{p^2}{q^3}, \tag{5.177b}$$

$$U_u = be^{-x}\frac{p}{q^2} + \lambda, \tag{5.177c}$$

$$U_x = -\lambda p, \tag{5.177d}$$

$$U_y = ae^{-x}\frac{p}{q} - \lambda q. \tag{5.177e}$$

Compatibility between these gives

$$\lambda_p + be^{-x}\frac{1}{q^2} = 0,$$ (5.178a)

$$p\lambda_p + \lambda - 2ae^{-x}\frac{p}{q^2} = 0,$$ (5.178b)

$$q\lambda_p - ae^{-x}\frac{1}{q} = 0,$$ (5.178c)

$$\lambda_q - 2be^{-x}\frac{p}{q^3} = 0,$$ (5.178d)

$$p\lambda_q + 2ae^{-x}\frac{p^2}{q^3} = 0,$$ (5.178e)

$$q\lambda_q + \lambda + ae^{-x}\frac{p}{q^2} = 0,$$ (5.178f)

$$\lambda_x + p\lambda_u - be^{-x}\frac{p}{q^2} = 0,$$ (5.178g)

$$\lambda_y + q\lambda_u = 0,$$ (5.178h)

$$q\lambda_x - p\lambda_y + ae^{-x}\frac{p}{q} = 0.$$ (5.178i)

From (5.178a) and (5.178c) we see that $b = -a$, and further from (5.178a) and (5.178b) we get

$$\lambda = ae^{-x}\frac{p}{q^2}.$$

This assignment shows that the entire system (5.178) is satisfied; thus, U exists satisfying the contact conditions. Integrating (5.177) (with $a = 1$) gives

$$U = e^{-x}\frac{p^2}{q^2},$$ (5.179)

and under the contact transformation

$$X = y + 2\frac{p}{q}, \quad Y = u, \quad U = e^{-x}\frac{p^2}{q^2},$$ (5.180)

gives U_{XY} as

$$U_{XY} = -\frac{e^{-x}\left(u_y^2 u_{xx} - 3u_x u_y u_{xy} + 2u_x^2 u_{yy} - u_x u_y^2\right)}{u_y\left(2z_y^2 z_{xx} - 4z_x z_y z_{xy} + 2z_x^2 z_{yy} - z_x u_z^2\right)} = 0.$$ (5.181)

Example 5.11 Consider

$$u_y(u_y + 1)u_{xx} - (2u_x u_y + u_x + u_y + 1)u_{xy} + u_x(u_x + 1)u_{yy} = 0.$$ (5.182)

This PDE admits the first integrals

$$F_1\left(x+u, \frac{q+1}{p}\right) = 0, \quad F_2\left(y+u, \frac{p+1}{q}\right) = 0. \tag{5.183}$$

Here we try

$$X = x+u, \quad Y = y+q, \tag{5.184a}$$

$$P = \frac{q+1}{p}, \quad Q = \frac{p+1}{q}. \tag{5.184b}$$

At this point, we need to find U such that the contact conditions (5.165) are satisfied, namely

$$U_p = 0,$$
$$U_q = 0,$$
$$U_u = \frac{p+1}{q} - \frac{q+1}{p} + \lambda, \tag{5.185}$$
$$U_x = \frac{q+1}{p} - \lambda p,$$
$$U_y = \frac{p+1}{q} - \lambda q.$$

However, cross differentiation of (5.185) shows they are not consistent; thus, another choice will be required. Next we try

$$X = x+u, \qquad Y = y+u, \tag{5.186a}$$

$$P = A\left(\frac{q+1}{p}\right), \quad Q = B\left(\frac{p+1}{q}\right), \tag{5.186b}$$

and determine whether the functions A and B can be chosen such that the contact conditions can be satisfied. With (5.186), the contact conditions (5.165) become

$$U_p = 0,$$
$$U_q = 0,$$
$$U_u = A\left(\frac{q+1}{p}\right) - B\left(\frac{p+1}{q}\right) + \lambda, \tag{5.187}$$
$$U_x = A\left(\frac{q+1}{p}\right) - \lambda p,$$
$$U_y = B\left(\frac{p+1}{q}\right) - \lambda q,$$

or, after eliminating λ, the latter two of (5.187) become

$$U_x + pU_u = (p+1)A\left(\frac{q+1}{p}\right) + pB\left(\frac{p+1}{q}\right), \tag{5.188a}$$

$$U_y + qU_u = qA\left(\frac{q+1}{p}\right) + (q+1)B\left(\frac{p+1}{q}\right). \tag{5.188b}$$

Differentiating (5.188a) with respect to q (or (5.188b) with respect to p) gives

$$q^2 A'\left(\frac{q+1}{p}\right) - p^2 B'\left(\frac{p+1}{q}\right) = 0. \tag{5.189}$$

If we let

$$p = \frac{s+1}{rs-1}, \quad q = \frac{r+1}{rs-1}, \tag{5.190}$$

then (5.189) becomes

$$(r+1)^2 A'(r) - (s+1)^2 B'(s) = 0, \tag{5.191}$$

from which we deduce that

$$(r+1)^2 A'(r) = (s+1)^2 B'(s) = -c_1 \tag{5.192}$$

for some constant c_1. Each can be integrated giving

$$A(r) = \frac{c_1}{r+1} + c_2, \quad B(s) = \frac{c_1}{s+1} + c_3. \tag{5.193}$$

Returning to (5.188) and simplifying gives

$$U_x + pU_u = (c_1 + c_2 + c_3)p + c_2, \tag{5.194a}$$

$$U_y + qU_u = (c_1 + c_2 + c_3)q + c_3. \tag{5.194b}$$

We choose $c_1 = 1$ and $c_2 = c_3 = 0$. Solving (5.194) subject to $U_p = U_q = 0$ gives $U = u$ and thus the contact transformation is

$$X = x + u, \quad Y = y + u, \quad U = u. \tag{5.195}$$

Under this contact (point) transformation, we have

$$U_{XY} = \frac{\left(u_y + u_y^2\right)u_{xx} - \left(2u_x u_y + u_x + u_y + 1\right)u_{xy} + \left(u_x + u_x^2\right)u_{yy}}{\left(1 + u_x + u_y\right)^3} = 0. \tag{5.196}$$

The reader might recall a similar problem in Example 5.9 and the results there could have been used here.

5.4.2   Parabolic MA Equations

As we've seen in the previous section, hyperbolic Monge-Ampere equations admitting four
first integrals can be linearized. In this section we consider parabolic Monge-Ampere equations admitting three first integrals. Suppose

$$Ar + Bs + Ct + D(rt - s^2) = E, \qquad (5.197)$$

which admits three general first integrals

$$F(\alpha, \beta, \gamma) = 0, \qquad (5.198)$$

where F is an arbitrary function of α, β and γ that are functions of x, y, u, p, and q. Then
it is possible to transform (5.197) to $U_{XX} = 0$.
We define

$$Y = \alpha, \quad P = \beta, \qquad (5.199)$$

and we will assume for the moment that X, U and Q exist such that these, in addition to
(5.199), satisfy the contact conditions (5.165). Consider

$$U_{XX} = 0.$$

This can be written as

$$\frac{\partial P}{\partial X} = \frac{\partial(P, Y)}{\partial(X, Y)} = \frac{\partial(P, Y)}{\partial(x, y)} \Big/ \frac{\partial(X, Y)}{\partial(x, y)} = 0,$$

giving

$$\frac{\partial(P, Y)}{\partial(x, y)} = 0, \qquad (5.200)$$

or

$$P_x Y_y - P_y Y_x = 0. \qquad (5.201)$$

The rest follows as in the hyperbolic case (steps (5.159)–(5.164)) and we refer the reader
there for details. However, unlike the hyperbolic case where there is a set of two independent
first integrals allowing one to choose X, Y, P, and Q, and use these to find U, in the parabolic
case we only have three first integrals and we have already used two to define Y and P. We
show how to use the third first integral to obtain X, U, and Q.
Recall the contact conditions:

$$U_p - PX_p - QY_p = 0,$$
$$U_q - PX_q - QY_q = 0,$$
$$U_u - PX_u - QY_u = \lambda, \tag{5.202}$$
$$U_x - PX_x - QY_x = -\lambda p,$$
$$U_y - PX_y - QY_y = -\lambda q.$$

We define U as

$$U = XP + W, \tag{5.203}$$

so that (5.202), after eliminating λ, becomes

$$W_p + XP_p - QY_p = 0, \tag{5.204a}$$
$$W_q + XP_q - QY_q = 0, \tag{5.204b}$$
$$W_x + pW_u + X(P_x + pP_u) - Q(Y_x + pY_u) = 0, \tag{5.204c}$$
$$W_y + qW_u + X(P_y + qP_u) - Q(Y_y + qY_u) = 0. \tag{5.204d}$$

Since Y and P are independent, then it is possible to choose a pair of equations from (5.204) and solve for X and Q. To demonstrate, let us suppose for the sake of argument that equations (5.204b) and (5.204c). From these two, we solve for X and Q

$$X = -\frac{(Y_x + pY_u)W_q - (W_x + pW_u)Y_q}{(Y_x + pY_u)P_q - (P_x + pP_u)Y_q}, \qquad Q = \frac{(W_x + pW_u)P_q - (P_x + pP_u)W_q}{(Y_x + pY_u)P_q - (P_x + pP_u)Y_q}, \tag{5.205}$$

noting that $(Y_x + pY_u)P_q - (P_x + pP_u)Y_q \neq 0$. Substituting (5.205) into the remaining equations in (5.204) gives

$$\frac{\partial(Y, P, W)}{\partial(\bar{x}, p, q)} = 0, \qquad \frac{\partial(Y, P, W)}{\partial(\bar{x}, \bar{y}, q)} = 0. \tag{5.206}$$

In (5.206) we have the usual Jacobian notation, and partial derivatives with respect to $\bar{x}$ and $\bar{y}$ are defined as

$$\frac{\partial}{\partial \bar{x}} = \frac{\partial}{\partial x} + p\frac{\partial}{\partial u}, \qquad \frac{\partial}{\partial \bar{y}} = \frac{\partial}{\partial y} + q\frac{\partial}{\partial u}. \tag{5.207}$$

As both Y and P are first integrals to (5.197), they both satisfy the equations

$$F_x + pF_u - \frac{C}{D}F_p + \frac{B}{2D}F_q = 0, \tag{5.208a}$$

$$F_y + qF_u + \frac{B}{2D}F_p - \frac{A}{D}F_q = 0. \tag{5.208b}$$

Using these in (5.206) gives rise to two equations for W, which is precisely (5.208). As we have a third independent solution of (5.208) (i.e. the third first integral), we use this for W. The following examples illustrate this method

Example 5.12 Consider

$$r + 2s + 2t + rt - s^2 = -1. \tag{5.209}$$

This equation admits the first integral

$$F(2x - y + p, \, y - x + q, \, u - xp - yq - x^2 + xy - y^2/2) = 0. \tag{5.210}$$

Here, we will choose

$$Y = y - x + q, \quad P = 2x - y + p. \tag{5.211}$$

The system (5.204) becomes

$$W_p + X = 0, \tag{5.212a}$$
$$W_q - Q = 0, \tag{5.212b}$$
$$W_x + pW_u + 2X + Q = 0, \tag{5.212c}$$
$$W_y + qW_u - X - Q = 0. \tag{5.212d}$$

We solve (5.212a) and (5.212b) for X and Q. This gives

$$X = -W_p, \quad Q = W_q, \tag{5.213}$$

giving the remaining two equations in (5.212) as

$$W_x + pW_u - 2W_p + W_q = 0, \tag{5.214a}$$
$$W_y + qW_u + W_p - W_q = 0. \tag{5.214b}$$

These are easily solved, giving (5.210). If we choose

$$W = u - xp - yq - x^2 + xy - y^2/2, \tag{5.215}$$

then from (5.213), (5.211), and (5.203) we have

$$X = x, \quad Y = y - x + u_y, \quad U = u - yu_y + x^2 - \frac{y^2}{2}, \tag{5.216}$$

and under this contact transformation

$$U_{XX} = \frac{u_{xx} + 2u_{xy} + 2u_{yy} + u_{xx}u_{yy} - u_{xy}^2 + 1}{u_{yy} + 1} = 0. \tag{5.217}$$

We also note that

$$x = Y, \quad y = -U_Y \quad u = U - X^2 - (X + Y)U_Y - \frac{1}{2}U_Y^2, \tag{5.218}$$

will transform (5.209) to $U_{XX} = 0$.

Example 5.13 Consider

$$xu_y^2 u_{xx} - 2xu_x u_y u_{xy} + xu_x^2 u_{yy} = u_x u_y^2 + xu_x^2 u_y. \tag{5.219}$$

This equation admits the first integral

$$F\left(u, x^2 + \frac{2xq}{p}, \frac{xq}{p} e^{-y}\right) = 0. \tag{5.220}$$

Here, we will choose

$$Y = u, \quad P = x^2 + \frac{2xq}{p}. \tag{5.221}$$

The system (5.204) becomes

$$W_p - \frac{2xq}{p} X = 0, \tag{5.222a}$$

$$W_q + \frac{2x}{p} X = 0, \tag{5.222b}$$

$$W_x + pW_u + 2\left(x + \frac{q}{p}\right) X - pQ = 0, \tag{5.222c}$$

$$W_y + qW_u - qQ = 0. \tag{5.222d}$$

We solve (5.222b) and (5.222d) for X and Q. This gives

$$X = -\frac{pW_q}{2x}, \quad Q = \frac{W_y + qW_u}{q}. \tag{5.223}$$

Thus, the remaining two equations in (5.222) become

$$pW_p + qW_q = 0, \tag{5.224a}$$

$$W_x - \frac{p}{q} W_y - \left(p + \frac{q}{x}\right) W_q = 0. \tag{5.224b}$$

These are easily solved, giving (5.220). If we choose

$$W = -\frac{2xqe^{-y}}{p}, \tag{5.225}$$

then from (5.223), (5.221), and (5.203) we have

$$X = e^{-y}, \quad Y = u, \quad U = x^2 e^{-y}, \tag{5.226}$$

and under this contact transformation

$$U_{XX} = \frac{xu_y^2 u_{xx} - 2xu_x u_y u_{xy} + xu_x^2 u_{yy} - u_x u_y^2 - xu_x^2 u_y}{e^{-y} u_x^3} = 0. \tag{5.227}$$

With the solution of (5.227) as $U = A(Y)X + B(Y)$, with A and B arbitrary, through the contact transformation (5.226) we obtain the general solution of the original PDE as

$$x^2 = A(u) + B(u)e^y. \tag{5.228}$$

Example 5.14 Consider

$$2uu_x u_y u_{xy} + u^2 \left(u_{xx} u_{yy} - u_{xy}^2 \right) = u_x^2 u_y^2. \tag{5.229}$$

This equation admits the first integral

$$F \left(x - \frac{u}{p}, y - \frac{u}{q}, \frac{u}{pq} \right) = 0. \tag{5.230}$$

Here, we will choose

$$Y = y - \frac{u}{q}, \quad P = \frac{u}{pq}. \tag{5.231}$$

The system (5.204) becomes

$$W_p - \frac{u}{p^2 q} X = 0, \tag{5.232a}$$

$$W_q - \frac{u}{pq^2} X - \frac{u}{q^2} Q = 0, \tag{5.232b}$$

$$W_x + p W_u + \frac{1}{q} X + \frac{p}{q} Q = 0, \tag{5.232c}$$

$$W_y + q W_u + \frac{1}{p} X = 0. \tag{5.232d}$$

We solve (5.232a) and (5.232b) for X and Q. This gives

$$X = -\frac{p^2 q W_p}{u}, \quad Q = -\frac{q(p W_p - q W_q)}{u}, \tag{5.233}$$

and the two remaining equations in (5.232) as

$$W_x + p W_u + \frac{pq}{u} W_q = 0, \tag{5.234a}$$

$$W_y + q W_u + \frac{pq}{u} W_p = 0. \tag{5.234b}$$

These are easily solved giving (5.230). If we choose

$$W = x - \frac{u}{p}, \tag{5.235}$$

then from (5.233), (5.231), and (5.203) we have

$$X = q, \quad Y = y - \frac{u}{q}, \quad U = x, \quad P = \frac{u}{pq}, \quad Q = -\frac{q}{p}, \tag{5.236}$$

and under this contact transformation

$$U_{XX} = \frac{2uu_x u_y u_{xy} + u^2 \left(u_{xx} u_{yy} - u_{xy}^2 \right) - u_x^2 u_y^2}{u_x^3 u_y^2 u_{yy}} = 0. \tag{5.237}$$

We note that through (5.236), we are able to explicitly solve for x, y, and u, giving

$$x = U, \quad y = Y - \frac{XP}{Q}, \quad u = -\frac{X^2 P}{Q}, \tag{5.238}$$

and one can show that under this transformation, (5.229) becomes $U_{XX} = 0$.

5.4.3 Elliptic MA Equations

We now extend our results to elliptic Monge-Ampere equations. We again consider

$$Ar + Bs + Ct + D(rt - s^2) = E, \tag{5.239}$$

and suppose that it admits two general first integrals, say

$$F\left(\alpha_1 \pm i\beta_1, \alpha_2 \pm i\beta_2 \right) = 0, \tag{5.240}$$

where F is an arbitrary function of $\alpha_i \pm i\beta_i$, $i = 1, 2$, and α_i, β_i are real functions of x, y, u, p, and q,. We will show it is possible to transform (5.239) to $U_{XX} + U_{YY} = 0$.
We define

$$X = \alpha_1, \quad Y = \beta_1, \quad P = \beta_2, \quad Q = \alpha_2 \tag{5.241}$$

and we'll assume that U exists such that the contact conditions are satisfied. Consider

$$U_{XX} + U_{YY} = 0.$$

This can be written as

$$\begin{aligned}
\frac{\partial P}{\partial X} + \frac{\partial Q}{\partial Y} &= \frac{\partial(P, Y)}{\partial(X, Y)} + \frac{\partial(X, Q)}{\partial(X, Y)} \\
&= \frac{\partial(P, Y)}{\partial(x, y)} \Big/ \frac{\partial(X, Y)}{\partial(x, y)} + \frac{\partial(X, Q)}{\partial(x, y)} \Big/ \frac{\partial(X, Y)}{\partial(x, y)} = 0,
\end{aligned} \tag{5.242}$$

giving

$$\frac{\partial(P, Y)}{\partial(x, y)} + \frac{\partial(X, Q)}{\partial(x, y)} = 0,$$

or

$$P_x Y_y - P_y Y_x + X_x Q_y - X_y Q_x = 0. \tag{5.243}$$

We now bring in (5.241), so (5.243) becomes

$$
\begin{aligned}
&\left(\beta_{2x} + p\beta_{2u} + r\beta_{2p} + s\beta_{2q}\right)\left(\beta_{1y} + q\beta_{1u} + s\beta_{1p} + t\beta_{1q}\right) \\
&- \left(\beta_{1x} + p\beta_{1u} + r\beta_{1p} + s\beta_{1q}\right)\left(\beta_{2y} + q\beta_{2u} + s\beta_{2p} + t\beta_{2q}\right) \\
&+ \left(\alpha_{1x} + p\alpha_{1u} + r\alpha_{1p} + s\alpha_{1q}\right)\left(\alpha_{2y} + q\alpha_{2u} + s\alpha_{2p} + t\alpha_{2q}\right) \\
&- \left(\alpha_{2x} + p\alpha_{2u} + r\alpha_{2p} + s\alpha_{2q}\right)\left(\alpha_{1y} + q\alpha_{1u} + s\alpha_{1p} + t\alpha_{1q}\right) = 0.
\end{aligned}
\tag{5.244}
$$

As F in (5.240) are first integrals, they satisfy the following (assuming that $D \neq 0$)

$$\alpha_{1x} - i\beta_{1x} + p\left(\alpha_{1u} - i\beta_{1u}\right) - \frac{C}{D}\left(\alpha_{1p} - i\beta_{1p}\right) + (\alpha_1 - i\beta_1)\left(\alpha_{1q} - i\beta_{1q}\right) = 0, \tag{5.245a}$$

$$\alpha_{1y} - i\beta_{1y} + q\left(\alpha_{1u} - i\beta_{1u}\right) + (\alpha_1 + i\beta_1)\left(\alpha_{1p} - i\beta_{1p}\right) - \frac{A}{D}\left(\alpha_{1q} - i\beta_{1q}\right) = 0, \tag{5.245b}$$

$$\alpha_{2x} - i\beta_{2x} + p\left(\alpha_{2u} - i\beta_{2u}\right) - \frac{C}{D}\left(\alpha_{2p} - i\beta_{2p}\right) + (\alpha_2 - i\beta_2)\left(\alpha_{2q} - i\beta_{2q}\right) = 0, \tag{5.245c}$$

$$\alpha_{2y} - i\beta_{2y} + q\left(\alpha_{2u} - i\beta_{2u}\right) + (\alpha_2 + i\beta_2)\left(\alpha_{2p} - i\beta_{2p}\right) - \frac{A}{D}\left(\alpha_{2q} - i\beta_{2q}\right) = 0, \tag{5.245d}$$

where

$$D^2\lambda^2 - BD\lambda + AC + DE = 0, \quad \lambda = a \pm ib. \tag{5.246}$$

We isolate real and imaginary parts of (5.245) and solve for $\alpha_{ix}, \alpha_{iy}, \beta_{ix}, \beta_{iy}$ and eliminate these in (5.244). This leads to

$$\frac{A}{D}r + 2as + \frac{C}{D}t + rt - s^2 = a^2 + b^2 - \frac{AC}{D^2}. \tag{5.247}$$

From (5.246), we have that

$$B - 2aD = 0, \tag{5.248a}$$

$$\left(a^2 - b^2\right)D^2 - aBD + AC + DE = 0, \tag{5.248b}$$

from which we deduce $\left(a^2 + b^2\right)D^2 = AC + DE$. Eliminating a and b in (5.247) gives

$$Ar + Bs + Ct + D(rt - s^2) = E,$$

which is (5.239). The next two examples illustrate these results.

Example 5.15 Consider

$$(1 + u_y^2)u_{xx} - 2u_x u_y u_{xy} + u_x^2 u_{yy} = 0. \tag{5.249}$$

As was shown in Example 5.3, this PDE admits the first integrals

$$F\left(u \pm iy, \frac{q \mp i}{p}\right) = 0. \tag{5.250}$$

Here we try

$$X = u, \quad Y = y, \quad P = \frac{1}{p}, \quad Q = \frac{q}{p}. \tag{5.251}$$

Unfortunately, the contact conditions are not satisfied, but with the slight adjustment,

$$X = u, \quad Y = y, \quad P = \frac{1}{p}, \quad Q = -\frac{q}{p}, \tag{5.252}$$

they lead to $U = x$ to which we add to (5.252) and under

$$X = u, \quad Y = y, \quad U = x, \quad P = \frac{1}{p}, \quad Q = -\frac{q}{p}. \tag{5.253}$$

or

$$x = U, \quad y = Y, \quad u = X, \quad p = \frac{1}{U_X}, \quad q = -\frac{U_Y}{U_X} \tag{5.254}$$

the PDE (5.249) is transformed to $U_{XX} + U_{YY} = 0$.

Example 5.16   Consider

$$u_{xx} + u_{yy} + 2u_{xx}u_{yy} - 2u_{xy}^2 = 0. \tag{5.255}$$

As was shown in Example 5.6, this PDE admits the first integrals

$$F(x + 2p \pm iy, q \pm ip) = 0. \tag{5.256}$$

Here we try

$$X = x + 2p, \quad Y = y, \quad P = p, \quad Q = q. \tag{5.257}$$

We find the contact conditions are satisfied and lead to $U = u + p^2$, and from

$$x = X - 2U_X, \quad y = Y, \quad u = U - U_X^2 \quad p = \frac{1}{U_X}, \quad q = -\frac{U_Y}{U_X}, \tag{5.258}$$

the PDE (5.255) is transformed to $U_{XX} + U_{YY} = 0$.

Example 5.17   Consider

$$u_y u_{xx} - u_{yy} + \left(1 + u_y\right)\left(u_{xx}u_{yy} - u_{xy}^2\right) = u_y. \tag{5.259}$$

This PDE admits the first integrals

$$F(p - x \pm iq, y + q + \ln q \pm ix) = 0. \tag{5.260}$$

Here we try

$$X = p - x, \quad Y = q, \quad P = x, \quad Q = y + q + \ln q. \tag{5.261}$$

We find that the contact conditions (5.165) are satisfied, leading to

$$U = xp + yq - u - \frac{1}{2}q^2 + q \ln q - q. \tag{5.262}$$

From (5.261) and (5.262)

$$x = U_X, \quad y = U_Y - Y - \ln Y, \quad u = X U_X + Y U_Y - U + \frac{1}{2}U_X^2 - \frac{1}{2}Y^2 - Y \tag{5.263}$$

transforms (5.259) to

$$U_{XX} + U_{YY} = 0. \tag{5.264}$$

5.5   Exercises

1. Find a first integral for the following

(i) $q^2 r - 2pqs + p^2 t = 0$,
(ii) $qr - (p + q + 1)s + (p + 1)t = 0$,
(iii) $rt - s^2 + 1 = 0$,
(iv) $yr - ps + t + y(rt - s^2) + 1 = 0$.

2. The equations that model one-dimensional gas flow are

$$\rho_t + (\rho u)_x = 0, \tag{5.265a}$$

$$u_t + u u_x = -\frac{P_x}{\rho}, \tag{5.265b}$$

where $P = P(\rho)$. If we introduce a stream function ψ such that

$$\rho = \psi_x, \quad \rho u = -\psi_t, \tag{5.266}$$

then (5.265a) is identically satisfied, whereas (5.265b) becomes

$$\psi_x^2 \psi_{tt} - 2\psi_t \psi_x \psi_{tx} + \psi_t^2 \psi_{xx} - \psi_x^2 P'(\psi_x) \psi_{xx} = 0. \tag{5.267}$$

Determine the forms of $P(\rho)$ such that (5.267) admits a first integral.

3. Show the PDE

$$2xu_y u_{xy} + u_y^2 u_{yy} + x^2 \left(u_{xx} u_{yy} - u_{xy}^2 \right) = u_y^2 \tag{5.268}$$

admits the first integral

$$F\left(\frac{q}{x}, u - xp - \frac{1}{2}q^2, p - \frac{yq + q^2}{x} \right) = 0. \tag{5.269}$$

Use this to find a contact transformation that linearizes the PDE.

4. Show the PDE

$$2xu_y u_{xy} + x^2 \left(u_{xx} u_{yy} - u_{xy}^2 \right) = 1 + u_y^2 \tag{5.270}$$

admits the first integral

$$F\left(\frac{1 \pm iq}{x}, y \pm i(xp - u) \right) = 0. \tag{5.271}$$

Use this to find a contact transformation that linearizes the PDE.

References

1. S. Lie, Neue Integrations methode der Monge-Ampkreshen Gleichung. Arch. Math. Kristiania **2**, 1–9 (1877)
2. A.R. Forsyth, *Theory of Differential Equations*, vol. 6 (Cambridge University Press, Cambridge, 1906)
3. M.H. Martin, The propagation of a plane shock into a quiet atmosphere. Can. J. Math. **5**, 37–39 (1953)
4. M.H. Martin, The Monge-Ampere partial differential equation $rt - s^2 + \lambda^2 = 0$. Pac. J. Math. **3**, 165–187 (1953)
5. S.V. Meleshko, *Methods for Constructing Exact Solutions of Partial Differential Equations* (Springer, 2005)
6. A.D. Polyanin, V.F. Zaitsev, *Handbook of Nonlinear Partial Differential Equations* (Chapman and Hall, CRC, Boca Raton, 2004)

Functional Separability

6

In a standard course in PDEs, one encounters the heat equation

$$u_t = u_{xx}, \tag{6.1}$$

and solutions in the form $u = T(t)X(x)$, often referred to as separable solutions. A natural question is: do nonlinear PDEs admit separable solutions? In general, the answer is no but sometimes they do admit some sort of a variation of separable solution. For example,

$$u_t = (uu_x)_x \tag{6.2}$$

does admit solutions of the form $u = TX$, leading to

$$\frac{T'}{T^2} = \frac{(XX')}{X}. \tag{6.3}$$

It also admits solutions of the form

$$u = a(t)x^2 + b(t)x + c(t). \tag{6.4}$$

Substitution of (6.4) into (6.2) and isolating coefficients of x gives

$$a' = 6a^2,$$

$$b' = 6ab, \tag{6.5}$$

$$c' = 2ac + b^2. \tag{6.6}$$

However

$$u_t = \frac{u_{xx}}{1 + u_x^2} \tag{6.7}$$

doesn't, since if $u = T(t)X(x)$, then

D. Arrigo, *Analytical Methods for Solving Nonlinear Partial Differential Equations*, Synthesis Lectures on Mathematics & Statistics. https://doi.org/10.1007/978-3-031-17069-0_6

$$T'X = \frac{TX''}{1 + T^2X'^2} \tag{6.8}$$

doesn't separate. However, the latter does admit separable solutions of the form $u = T(t) + X(x)$.

Sometimes it is not at all obvious that a particular PDE admits some form of a separable solution. The following two examples illustrate this.

Example 6.1 Consider

$$u_{tt} = u_{xx} + e^u \tag{6.9}$$

and separable solutions of the form

$$u = \ln |T(t) + X(x)|. \tag{6.10}$$

Substitution of (6.10) into (6.9) gives

$$\frac{(T + X)T'' - T'^2}{(T + X)^2} = \frac{(T + X)X'' - X'^2}{(T + X)^2} + (T + X). \tag{6.11}$$

Simplifying gives

$$(T + X)T'' - T'^2 = (T + X)X'' - X'^2 + (T + X)^3, \tag{6.12}$$

and differentiating (6.12) with respect to t and x gives

$$X'T''' = T'X''' + 6a(T + X)T'X', \tag{6.13}$$

or

$$\frac{T'''}{T'} = \frac{X'''}{X'} + 6(T + X). \tag{6.14}$$

This leads to the two equations

$$T''' - 6TT' = kT', \quad X''' + 6XX' = kX', \tag{6.15}$$

where k is an arbitrary constant sometimes refereed to as a separation constant. Integrating each twice gives

$$\frac{1}{2}T'^2 - T^3 = \frac{1}{2}kT^2 + c_1T + c_2, \quad \frac{1}{2}X'^2 + X^3 = \frac{1}{2}kX^2 + c_3X + c_4, \tag{6.16}$$

where $c_1 - c_4$ are constants of integration. Finally, substituting (6.16) back into (6.12) shows that it is identically satisfied if $c_1 + c_3 = 0$ and $c_2 - c_4 = 0$. Thus, any solution of

$$T'^2 = 2T^3 + kT^2 - 2c_1T + 2c_2, \quad X'^2 = -2X^3 + kX^2 + 2c_1X + 2c_2, \tag{6.17}$$

when substituted into (6.10), will give rise to an exact solution of the PDE (6.9).

Example 6.2 Consider

$$u_{tt} - u_{xx} = \sin u, \tag{6.18}$$

and solutions of the form

$$u = 4\tan^{-1}\left(\frac{X(x)}{T(t)}\right). \tag{6.19}$$

Substituting (6.19) into (6.18) and re-arranging terms gives

$$2T'^2 - TT'' - X^2\frac{T''}{T} + 2X'^2 - T^2\frac{X''}{X} - XX'' - T^2 - X^2 = 0. \tag{6.20}$$

Differentiating (6.20) twice with respect to t and x gives

$$\frac{1}{TT'}\left(\frac{T''}{T}\right)' + \frac{1}{XX'}\left(\frac{X''}{X}\right)' = 0, \tag{6.21}$$

leading to the equations

$$\frac{1}{TT'}\left(\frac{T''}{T}\right)' = k, \quad \frac{1}{XX'}\left(\frac{X''}{X}\right)' = -k, \tag{6.22}$$

where k is again a constant. Each can be integrated twice, leading to

$$X'^2 = c_1 X^4 + c_2 X^2 + c_3, \quad T'^2 = -c_1 T^4 + c_4 T^2 + c_5. \tag{6.23}$$

Substituting (6.23) into (6.20) leads to the conditions $c_4 - c_2 = 1, c_3 + c_5 = 0$. Any solution of the Jacobi elliptic equations (6.23) with these constant restrictions will lead to an exact solution of (6.18) in the form of (6.19).

A natural question is: how did one know that separable solutions of

$$u_{tt} = u_{xx} + e^u, \tag{6.24}$$

were of the form

$$u = \ln|T(t) + X(x)|, \tag{6.25}$$

or separable solutions of

$$u_{tt} - u_{xx} = \sin u, \tag{6.26}$$

were of the form

$$u = 4\tan^{-1}\left(\frac{X(x)}{T(t)}\right)? \tag{6.27}$$

We also ask: are there others? If an equation admits functional separable solutions, then it would be of the form

$$f(u) = T(t)X(x), \quad \text{or} \quad f(u) = T(t) + X(x).$$

But these are equivalent, as we can take natural logarithms of both sides of the first and reset f, T and X. Thus, we only consider

$$f(u) = T(t) + X(x). \tag{6.28}$$

Differentiating (6.28) with respect to t and x gives

$$f'u_{tx} + f''u_t u_x = 0,$$

or

$$u_{tx} = F(u)u_t u_x, \tag{6.29}$$

where $F = -f''/f'$. If a PDE admits solutions of the form (6.28) then it will be compatible with (6.29). As an example, we will reconsider (6.9). Taking t and x derivatives of both (6.9) and (6.29) gives

$$u_{ttt} - u_{txx} = e^u u_t, \tag{6.30a}$$

$$u_{ttx} - u_{xxx} = e^u u_x, \tag{6.30b}$$

$$u_{ttx} = F'u_t^2 u_x + Fu_{tt}u_x + Fu_t u_{tx}, \tag{6.30c}$$

$$u_{txx} = F'u_t u_x^2 + Fu_{tx}u_x + Fu_t u_{xx}. \tag{6.30d}$$

Solving (6.30) for the third order derivatives $u_{ttt}, u_{ttx}, u_{txx}$, and u_{xxx} gives

$$u_{ttt} = Fu_x u_{tx} + Fu_t u_{xx} + F'u_t u_x^2 + e^u u_t \tag{6.31a}$$

$$u_{ttx} = Fu_x u_{tt} + Fu_t u_{tx} + F'u_t^2 u_x \tag{6.31b}$$

$$u_{txx} = Fu_x u_{tx} + Fu_t u_{xx} + F'u_t u_x^2 \tag{6.31c}$$

$$u_{xxx} = Fu_x u_{tt} + Fu_t u_{tx} + F'u_t^2 u_x - e^u u_x. \tag{6.31d}$$

Requiring compatibility of (6.31)

$$(u_{ttx})_t = (u_{ttt})_x, \quad (u_{txx})_t = (u_{ttx})_x, \quad (u_{xxx})_t = (u_{txx})_x, \tag{6.32}$$

and using both (6.9) and (6.29) gives

$$\left(F'' + 2FF'\right)\left(u_t^3 u_x + u_x^3\right) + \left(3F' + 2F^2 + F - 1\right)e^u u_t u_x = 0, \tag{6.33}$$

from which we obtain two equations for F

$$F'' + 2FF' = 0, \tag{6.34a}$$

$$3F' + 2F^2 + F - 1 = 0. \tag{6.34b}$$

Requiring that (6.34) be compatible gives

$$F = -1, \quad F = \frac{1}{2}. \tag{6.35}$$

Since $F = -f''/f'$, upon integrating we get

$$f = c_1 e^u + c_2, \quad f = c_1 e^{-u/2} + c_2, \tag{6.36}$$

where c_1 and c_2 are arbitrary constants. Thus,

$$u_{tt} = u_{xx} + e^u \tag{6.37}$$

admits separable solutions of the form

$$c_1 e^u + c_2 = T(t) + X(x), \tag{6.38a}$$
$$c_1 e^{-u/2} + c_2 = T(t) + X(x). \tag{6.38b}$$

Note that we can set $c_1 = 1$ and $c_2 = 0$ without loss of generality. Solving for u, we obtain

$$u = \ln |T(t) + X(x)|, \tag{6.39a}$$
$$u = -2 \ln |T(t) + X(x)|, \tag{6.39b}$$

and the first solution in (6.39) we saw earlier.

Example 6.3 Seek functional separable solutions to the nonlinear diffusion equation

$$u_t = \left(\frac{u_x}{u^2}\right)_x. \tag{6.40}$$

We now wish that (6.40) and

$$u_{tx} = F(u)u_t u_x, \tag{6.41}$$

are compatible. From our original PDE (6.40), we have

$$u_{xx} = u^2 u_t + \frac{2u_x^2}{u}. \tag{6.42}$$

From cross differentiation $(u_{xx})_t = (u_{tx})_t$ and (6.42) and (6.41), we obtain

$$u^2 u_{tt} + 2uu_t^2 + \frac{4u_x u_{tx}}{u} - \frac{2u_x^2 u_t}{u^2} = F' u_t u_x^2 + F u_x u_{tx} + F u_t u_{xx}. \tag{6.43}$$

Using (6.42) and (6.41) to eliminate derivatives u_{tx} and u_{xx}, we obtain

$$u_{tt} = \frac{F' u_t u_x^2}{u^2} - \frac{2u_t^2}{u} + \frac{2u_t u_x^2}{u^4} + \frac{F^2 u_t u_x^2}{u^2} + F u_t^2 - \frac{2F u_t u_x^2}{u^3}. \tag{6.44}$$

We now further require that (6.41) and (6.44) be compatible. This leads to

$$2\left(F' + F^2 - \frac{3F}{u} + \frac{3}{u^2}\right)u_t^2 u_x + \left(\frac{F''}{u^2} + \frac{2FF'}{u^2} + \frac{2F^2}{u^3} - \frac{2F}{u^4}\right)u_t u_x^3 = 0, \quad (6.45)$$

from which we obtain two equations for F

$$F' + F^2 - \frac{3F}{u} + \frac{3}{u^2} = 0, \tag{6.46a}$$

$$\frac{F''}{u^2} + \frac{2FF'}{u^2} + \frac{2F^2}{u^3} - \frac{2F}{u^4} = 0. \tag{6.46b}$$

Requiring that (6.46) be compatible gives

$$F = \frac{1}{u}, \quad F = \frac{3}{u}.$$

Each case will be considered separately.
Case 1: $F = u^{-1}$. Since $F = -f''/f'$, this gives

$$\frac{f''}{f'} = -\frac{1}{u},$$

and integrating twice gives

$$f(u) = c_1 + c_2 \ln u.$$

Thus, our equation admits a separable solution of the form

$$\ln u = T(t) + X(x).$$

Note that we set $c_1 = 1$ and $c_2 = 0$ without loss of generality. With a resetting of variables, $e^T \to T, e^X \to X$, the separable solution is just $u = T(t)X(x)$.

Case 2: $F = u^{-3}$. Since $F = -f''/f'$, this gives

$$\frac{f''}{f'} = -\frac{3}{u},$$

and integrating twice gives

$$f(u) = c_1 + \frac{c_2}{u^2}.$$

Thus, our equation admits a separable solution of the form

$$u = \frac{1}{\sqrt{T(t) + X(x)}}. \tag{6.47}$$

Note that we set $c_1 = 1$ and $c_2 = 0$ without loss of generality. Substituting (6.47) into our original PDE shows that

$$2XX'' - X'^2 + 2TX'' - 2T' = 0,$$

which eventually leads to the exact solution

$$u = \frac{1}{\sqrt{c_2 x^2 + c_1 x + c_0 + c_3 e^{2c_2 t}}}.$$

General Traveling Wave Solutions

We now wish to consider a slightly different form of solutions, namely

$$f(u) = A(t)x + B(t), \tag{6.48}$$

which are known as general traveling wave solutions. As we did previously, we differentiate (6.48) , but this time with respect to x twice, giving

$$f'u_{xx} + f''u_x^2 = 0, \tag{6.49}$$

or

$$u_{xx} = F(u)u_x^2, \tag{6.50}$$

where $F = -f''/f'$. If our given equation is compatible with (6.50), then it will admit solutions in the form (6.48). The following examples illustrate this.

Example 6.4 Consider the diffusion equation with a nonlinear source

$$u_t = u_{xx} + Q(u). \tag{6.51}$$

Using (6.50), (6.51) becomes

$$u_t = F(u)u_x^2 + Q(u). \tag{6.52}$$

Imposing compatibility between (6.50) and (6.52) gives

$$Q'' - FQ' - QF'' + \left(F'' + 4FF' + 2F^3\right)u_x^2 = 0, \tag{6.53}$$

from which we obtain the determining equations

$$Q'' - FQ' - QF'' = 0, \tag{6.54a}$$
$$F'' + 4FF' + 2F^3 = 0. \tag{6.54b}$$

Since $F = -f''/f'$, then (6.54a) integrates readily, giving

$$Q = \frac{c_1 f + c_2}{f'}, \tag{6.55}$$

where c_1 and c_2 are arbitrary constants. The remaining equation in (6.54) becomes

$$\frac{f^{(4)}}{f'} - \frac{7 f'' f'''}{f'^2} + \frac{8 f''^3}{f'^3} = 0.$$
(6.56)

Fortunately, (6.56) admits several integrating factors

$$\frac{1}{f'^2}, \quad \frac{f}{f'^2}, \quad \frac{f^2}{f'^2}, \quad \frac{\ln f'}{f'^2},$$
(6.57)

and can be used to integrate (6.56). The first integrating factor leads to

$$\frac{f'''}{f'^3} - 2\frac{f''^2}{f'^4} = c_3,$$
(6.58)

the integrating factor f' leads to

$$\frac{f''}{f'^2} = c_3 f + c_4,$$
(6.59)

and the integrating factor f' leads to

$$\ln|f'| = \frac{1}{2} c_3 f^2 + c_4 f + c_5,$$
(6.60)

or

$$f' = e^{c_3 f^2 + c_4 f + c_5},$$
(6.61)

where $c_3 - c_5$ are further arbitrary constant. Note that we have absorbed the factor of $1/2$ into c_3. For example, if we choose $c_3 = 0, c_4 = -1, c_5 = 0$ then (6.61) integrates to give

$$f = \ln u,$$
(6.62)

where we have suppressed the constant of integration. In this case, Q in (6.55) becomes

$$Q = (c_1 \ln u + c_2)u,$$
(6.63)

and PDEs of the form

$$u_t = u_{xx} + (c_1 \ln u + c_2)u$$
(6.64)

admits solutions of the form

$$u = e^{A(t)x + B(t)}.$$
(6.65)

The reader can verify that substituting (6.65) into (6.64) shows that A and B satisfy the ODE

$$A' = c_1 A, \quad B' = -2A^2 + c_1 B + c_2.$$
(6.66)

Example 6.5 Consider the nonlinear diffusion equation

$$u_t = (D(u)u_x)_x \, . \tag{6.67}$$

Expanding (6.67) and using (6.50) gives

$$u_t = (D' + DF)u_x^2. \tag{6.68}$$

If we set $G = D' + DF$, then (6.68) becomes

$$u_t = Gu_x^2. \tag{6.69}$$

Requiring that (6.50) and (6.69) be compatible gives

$$G'' + 3FG' + (F' + 2F^2)G = 0, \tag{6.70}$$

and since $F = -f''/f'$, then (6.70) becomes

$$G'' + 3\frac{f''}{f'}G' - \left(\frac{f'''}{f'} - 3\frac{f''^2}{f'^2}\right)G = 0. \tag{6.71}$$

Equation (6.71) can be explicitly solved, giving

$$G = (c_1 f + c_2)\, f', \tag{6.72}$$

where c_1 and c_2 are constants of integration. Further, since $G = D' + FD$, then

$$D' - \frac{f''}{f'}D = (c_1 f + c_2)\, f'. \tag{6.73}$$

Since $f(u) = A(t)x + B(t)$ we have the freedom to set the constants c_1 and c_2 without loss of generality. If $c_1 = 0$, then we can set $c_2 = 1$ without loss of generality. If $c_1 \neq 0$, we can set $c_1 = \pm 1$ and $c_2 = 0$. In the first case, where $c_1 = 0$, we can solve (6.73) giving

$$f(u) = \int \frac{D(u)}{u+d}\, du = A(t)x + B(t), \tag{6.74}$$

(d is an arbitrary constant) and substitution of (6.74) into (6.67) gives

$$A' = 0, \quad B' = A^2. \tag{6.75}$$

These easily integrate, giving

$$\int \frac{D(u)}{u+d}\, du = c_1 x + c_1^2 t + c_2, \tag{6.76}$$

is an exact solution to the nonlinear PDE

$$u_t = (D(u)u_x)_x \, . \tag{6.77}$$

Table 6.1 Exact solutions of the nonlinear diffusion equation (6.67)

$f(u)$	Solution	$D(u)$
u	$u = \dfrac{x}{\sqrt{2t}}$	$d - \frac{1}{2}u^2$
u^2	$u = \pm\sqrt{\dfrac{x}{\sqrt{2t}}}$	$2u(d - \frac{1}{3}u^3)$
$\dfrac{1}{u}$	$u = \dfrac{\sqrt{2t}}{x}$	$\dfrac{\ln|u| - d}{u^2}$
e^u	$u = \ln\dfrac{x}{\sqrt{2t}}$	$(d - e^u)e^u$
$\tan u$	$u = \tan^{-1}\left(\dfrac{x}{\sqrt{2t}}\right)$	$(d + \ln|\cos u|)\sec^2 u$

Of course, (6.76) is nothing more than the usual traveling solution.

In the second case we consider $c_1 = -1$, (6.73) becomes

$$D' - \frac{f''}{f'}D = -ff'. \tag{6.78}$$

Substitution of $f(u) = A(t)x + B(t)$ into (6.67), with D and F satisfying (6.78), leads to

$$A' = -A^3, \quad B' = -A^2 B. \tag{6.79}$$

These easily integrate

$$A = \frac{1}{\sqrt{2(t + t_0)}}, \quad B = \frac{x_0}{\sqrt{2(t + t_0)}}, \tag{6.80}$$

and we set $t_0 = x_0 = 0$ without loss of generality. Thus, we obtain solutions of the form

$$f(u) = \frac{x}{\sqrt{2t}}. \tag{6.81}$$

Now (6.78) would need to be integrated for a given D. However, this is a nonlinear ODE for F. Instead, for a given F, we will deduce a form of D, where exact solutions (6.67) can be obtained. Integrating (6.78) gives

$$D(u) = \left(d - \int f(u)\,du\right)f'(u), \tag{6.82}$$

where d is a constant of integration. Table 6.1 contains several examples where exact solutions of the nonlinear diffusion equation (6.67) are given for a variety of diffusivities.

Example 6.6 Consider the nonlinear dispersion equation

$$u_t + k(u)u_x + u_{xxx} = 0, \tag{6.83}$$

where $k(u)$ is an arbitrary function of u. Our goal is to determine forms of k such that (6.83) admits separable solutions. Expanding (6.83) and using (6.50) gives

$$u_t = -k(u)u_x - \left(F' + 2F^2\right)u_x^3. \tag{6.84}$$

Requiring that (6.84) and (6.50) be compatible gives rise to the following set of determining equations

$$k'' + Fk' = 0, \tag{6.85a}$$
$$F''' + 9FF'' + 6F'^2 + 30F^2F' = 0. \tag{6.85b}$$

Since $F = -f''/f'$, then (6.85a) integrates to give

$$k = c_1 f + c_2, \tag{6.86}$$

where c_1 and c_2 are arbitrary constants. Furthermore, (6.85b) becomes

$$-\frac{f^{(5)}}{f'} + \frac{13f''f^{(4)}}{f'^2} + \frac{9f'''^2}{f'^2} - \frac{81f''^2f'''}{f'^3} + \frac{72f''^4}{f'^4} = 0, \tag{6.87}$$

which possesses the integrating factor f'^{-3}, which allows us to integrate, giving

$$\frac{f^{(4)}}{f'^4} - \frac{9f'''f''}{f'^5} + \frac{12f''^3}{f'^6} + c_3 = 0. \tag{6.88}$$

Equation (6.88) possesses the integrating factor f' leading to

$$\frac{f'''}{f'^3} - \frac{3f''^2}{f'^4} + c_3 f + c_4 = 0, \tag{6.89}$$

where c_3 and c_4 are additional arbitrary constants. If we perform a Hodograph transformation on (6.89) (interchanging the roles of f and u) we get the linear ODE

$$u_{fff} - (c_3 f + c_4) u_f = 0. \tag{6.90}$$

If $c_3 \neq 0$ this can be solved in terms of integrated Airy functions. If $c_3 = 0$, then (6.89) can be integrated twice, giving

$$f' = \frac{1}{\sqrt{c_4 u^2 + c_5 u + c_6}}. \tag{6.91}$$

It can further be integrated, depending on the sign of c_4. Here we consider the cases where $c_4 = -1$, $c_4 = 0$, and $c_4 = 1$. We will also set $c_1 = 1$, $c_2 = 0$.

Case 1. $c_4 = -1$ In this case, (6.91) integrates to give

$$f = \tan^{-1}\left(\frac{u - \frac{1}{2}c_5}{\sqrt{-u^2 + c_5 u + c_6}}\right) + c_7. \tag{6.92}$$

We will set $c_2 = c_5 = c_7 = 0$ and $c_1 = c_6 = 1$. From (6.83), (6.86), and (6.48), PDEs of the form

$$u_t + \tan^{-1}\left(\frac{u}{\sqrt{1-u^2}}\right)u_x + u_{xxx} = 0 \tag{6.93}$$

admit separable solutions of the form

$$u = \sin\left(A(t)x + B(t)\right), \tag{6.94}$$

where A and B satisfy

$$A' + A^2 = 0, \quad B' + AB - A^3 = 0. \tag{6.95}$$

These are easily solved, giving

$$A = \frac{1}{t + a_0}, \quad B = \frac{b_0}{t + a_0} - \frac{1}{(t + a_0)^2}, \tag{6.96}$$

a_0, b_0 are constant, and via (6.94) (with $a_0 = b_0 = 0$) gives

$$u = \sin\left(\frac{xt - 1}{t^2}\right). \tag{6.97}$$

Case 2. $c_4 = 0$ In this case (6.91) integrates to give

$$f = \frac{2\sqrt{c_5 u + c_6}}{c_5} + c_7. \tag{6.98}$$

We will set $c_1 = 1$, $c_2 = 0$, $c_5 = 4$, $c_6 = 0$, $c_7 = 0$. From (6.83), (6.86), and (6.48), PDEs of the form

$$u_t + \sqrt{u}\, u_x + u_{xxx} = 0 \tag{6.99}$$

admit separable solutions of the form

$$u = (A(t)x + B(t))^2, \tag{6.100}$$

where A and B satisfy

$$A' + A^2 = 0, \quad B' + AB = 0. \tag{6.101}$$

These are easily solved, giving

$$A = \frac{1}{t + a_0}, \quad B = \frac{b_0}{t + a_0}, \tag{6.102}$$

a_0, b_0 are constant, and via (6.100) (with $a_0 = b_0 = 0$) gives

$$u = \left(\frac{x}{t}\right)^2.$$

(6.103)

Case 3. $c_4 = 1$ In this case, (6.91) integrates to give

$$f = \ln\left|\frac{1}{2}c_5 + u + \sqrt{u^2 + c_5 u + c_6}\right| + c_7.$$

(6.104)

We will set $c_2 = c_5 = c_7 = 0$ and $c_1 = c_6 = 1$. From (6.83), (6.86), and (6.48), PDEs of the form

$$u_t + \ln\left|u + \sqrt{u^2 + 1}\right| u_x + u_{xxx} = 0$$

(6.105)

admit separable solutions of the form

$$u = \sinh\left(A(t)x + B(t)\right),$$

(6.106)

where A and B satisfy

$$A' + A^2 = 0, \quad B' + AB + A^3 = 0.$$

(6.107)

These are easily solved, giving

$$A = \frac{1}{t + a_0}, \quad B = \frac{b_0}{t + a_0} + \frac{1}{(t + a_0)^2},$$

(6.108)

a_0, b_0 are constant, and via (6.106) (with $a_0 = b_0 = 0$) gives

$$u = \sinh\left(\frac{xt + 1}{t^2}\right).$$

(6.109)

If we were to chose $c_6 = -1$, we would obtain the exact solution

$$u = \cosh\left(\frac{xt + 1}{t^2}\right)$$

(6.110)

for the PDE

$$u_t + \ln\left|u + \sqrt{u^2 - 1}\right| u_x + u_{xxx} = 0.$$

(6.111)

Higher Dimensional Functional Separable Solutions

In this section we extend our results to PDEs in 3 dimensions, and, in particular, we consider the nonlinear elliptic equation

$$u_{xx} + u_{yy} + u_{zz} = Q(u).$$

(6.112)

We will seek separable solutions of the form

$$f(u) = A(x) + B(y) + C(z).$$ (6.113)

The reader can verify that (6.113) arises on solving

$$u_{xy} = F(u)u_x u_y, \quad u_{xz} = F(u)u_x u_z, \quad u_{yz} = F(u)u_y u_z,$$ (6.114)

where $F = -f''/f'$. We wish to make (6.112) and (6.114) compatible. By differentiating (6.112) and (6.114) with respect to x, y, and z, we calculate all third order derivatives. Requiring that these be compatible with each other gives rise to

$$2(F' + F^2)(u_{xx} - Fu_x^2) - (F'' + 2FF')(u_x^2 + u_y^2 + u_z^2) + Q'' - FQ' - (3F' + 2F^2)Q = 0,$$ (6.115a)

$$2(F' + F^2)(u_{yy} - Fu_y^2) - (F'' + 2FF')(u_x^2 + u_y^2 + u_z^2) + Q'' - FQ' - (3F' + 2F^2)Q = 0,$$ (6.115b)

$$2(F' + F^2)(u_{zz} - Fu_z^2) - (F'' + 2FF')(u_x^2 + u_y^2 + u_z^2) + Q'' - FQ' - (3F' + 2F^2)Q = 0,$$ (6.115c)

and the first order condition

$$\left(3F'' + 8FF' + 2F^3\right)\left(u_x^2 + u_y^2 + u_z^2\right) - 3Q'' + 3FQ' + (7F' + 4F^2)Q = 0.$$ (6.116)

We can solve (6.115) for u_{xx}, u_{yy}, and u_{zz}, provided that $F' + F^2 \neq 0$. As $F' + F^2 = 0$ is a special case, we consider this first. This gives $F = u^{-1}$, with the constant of integration omitted. Equation (6.115) reduce to

$$u^2 Q'' - uQ' + Q = 0.$$ (6.117)

This integrates to

$$Q = q_1 u + q_2 u \ln |u|,$$ (6.118)

where q_1 and q_2 are arbitrary constants, with $q_2 \neq 0$ as we are interested in nonlinear PDEs. However, as we can scale u in our original PDE (6.112) with this particular Q, we can set $q_1 = 0$ without loss of generality. Since $F = -f''/f'$, then from (6.113) we obtain

$$u = e^{A+B+C}.$$ (6.119)

Substitution of (6.119) into (6.112) with (6.118) gives

$$\left(A'' + A'^2 + B'' + B'^2 + C'' + C'^2\right)e^{A+B+C} = q_2 e^{A+B+C}(A + B + C),$$ (6.120)

which we see separates into

$$A'' + A'^2 = q_2 A + a, \quad B'' + B'^2 = q_2 B + b, \quad C'' + C'^2 = q_2 C + c,$$ (6.121)

where a, b and c are constants such that $a + b + c = 0$. We can further transfer the constants a, b, and c to the solution (6.119), allowing us to set $a = b = c = 0$ and the ODEs in (6.121) become the single ODE $w'' + w'^2 = q_2 w$, $w = w(\xi)$ which can be solved to quadrature. If the first constant of integration is omitted, we obtain the explicit solution

$$w = \frac{1}{2} + \frac{q_2}{4}(\xi - w_0)^2, \tag{6.122}$$

where w_0 is a constant of integration. Thus, the forms of A, B, and C are

$$A = \frac{1}{2} + \frac{q_2}{4}(x - x_0)^2, \quad B = \frac{1}{2} + \frac{q_2}{4}(y - y_0)^2, \quad C = \frac{1}{2} + \frac{q_2}{4}(z - z_0)^2. \tag{6.123}$$

As we have the flexibility of translational and scaling symmetry, we arrive at our final solution

$$u = e^{q(x^2 + y^2 + z^2) + 3/2} \tag{6.124}$$

as an exact solution to the PDE

$$u_{xx} + u_{yy} + u_{zz} = 4qu \ln |u|. \tag{6.125}$$

Note that we set $q_2 = 4q$.

We now assume that $F' + F^2 \neq 0$. We now can solve for all second order derivatives. From (6.115) we obtain

$$u_{xx} = Fu_x^2 + \frac{(F'' + 2FF')(u_x^2 + u_y^2 + u_z^2) + Q'' - FQ' - (3F' + 2F^2)Q}{2(F' + F^2)}, \tag{6.126a}$$

$$u_{yy} = Fu_y^2 + \frac{(F'' + 2FF')(u_x^2 + u_y^2 + u_z^2) + Q'' - FQ' - (3F' + 2F^2)Q}{2(F' + F^2)}, \tag{6.126b}$$

$$u_{zz} = Fu_z^2 + \frac{(F'' + 2FF')(u_x^2 + u_y^2 + u_z^2) + Q'' - FQ' - (3F' + 2F^2)Q}{2(F' + F^2)}. \tag{6.126c}$$

to which we add those given in (6.114). Requiring (6.114) and (6.126) be compatible gives

$$\left(F''' + 3FF'' + 2F'^2 + 2F^2F'\right)\left(u_x^2 + u_y^2 + u_z^2\right)$$
$$- Q''' + 2FQ'' + (4F' + F^2)Q' + (3F'' + FF' - 2F^3)Q = 0. \tag{6.127}$$

Eliminating $u_x^2 + u_y^2 + u_z^2$ between (6.116) and (6.127) gives rise to our first constraint on F and Q.

$$\left(3F'' + 8FF' + 2F^3\right)\left(-Q''' + 2FQ'' + (3F' + F^2)Q' + (3F'' + FF' - 2F^3)Q\right)$$
$$- \left(F''' + 3FF'' + 2F'^2 + 2F^2F'\right)\left(-3Q'' + 3FQ' + (7F' + 4F^2)Q\right) = 0. \tag{6.128}$$

If we differentiate (6.116) with respect to either x, y, or z and use (6.114), (6.116), and (6.126) we obtain a second constraint

$$3(3F'' + 8FF' + 2F^3)Q''' - (9F''' + 45FF'' + 24F'^2 + 74F^2F')Q''$$
$$+(9FF''' - 30F'F'' + 24F^2F'' - 56FF'^2 - 2F^3F')Q' + (21F' + 12F^2)F''' - 27F''^2$$
$$+ (26F^3 - 28FF')F'' + 56F'^3 + 42F^2F'^2 + 48F^4F' + 8F^6 = 0. \qquad (6.129)$$

Eliminating Q''' between (6.128) and (6.129) gives two factors: one is $F' + F^2$, which we assume is not zero; the second is

$$(3F' + F^2)Q'' - (3F'' + 11FF' + 3F^3)Q' + (5FF'' - 7F'^2 + 7F^2F' + 2F^4)Q = 0. \qquad (6.130)$$

It is left as an exercise to the reader to show that if (6.130) is satisfied, then (6.128) and (6.129) are automatically satisfied. To solve (6.130) for a given Q would seem like a daunting task — solving a highly nonlinear second order ODE for F. However, the ODE is linear for Q and it is this road we will take. We will first introduce the substitution $F = 3S'/S$. The motivation for this is in the first term $3F' + F^2$. This gives

$$S^2S''Q'' - (S^2S''' + 8SS'S'')Q' + (5SS'S''' - 7SS''^2 + 20S'^2S'')Q = 0. \qquad (6.131)$$

One exact solution of (6.131) would then allow us to reduce the order of the ODE. Seeking a solution of the form $Q = Q(S, S')$, we are lead to the solution $Q = S^5S'$; letting $Q = S^5S'P$ where $P = P(u)$ gives

$$SS'S''P'' - (SS'S''' - 2SS''^2 - 2S'^2S'')P' = 0, \qquad (6.132)$$

from which we integrate, giving

$$P = p_1 + p_2 \int \frac{S''}{S^2S''^2}du, \qquad (6.133)$$

where p_1 and p_2 are constant. Now we will relate this to f. Since $F = \dfrac{3S'}{S} = -\dfrac{f''}{f'}$, then $S^3 = \dfrac{1}{f'}$, which allows (6.133) to be integrated, giving

$$P = p_1 + p_2 \left(3\frac{f'^2}{f''} - 2f\right). \qquad (6.134)$$

We note that $f'' \neq 0$ as $f'' = 0$ gives $Q = 0$. This, in turn gives

$$Q = -\frac{p_1}{3}\frac{f''}{f'^3} - \frac{p_2}{3}\frac{f''}{f'^3}\left(3\frac{f'^2}{f''} - 2f\right), \qquad (6.135)$$

or

$$Q = q_1 \frac{f''}{f'^3} + q_2 \left(\frac{2ff''}{f'^3} - \frac{3}{f'} \right), \tag{6.136}$$

where q_1 and q_2 are constant. Now that we know the form of Q, we go after the forms of A, B, and C. Recall that $f(u) = A(x) + B(y) + C(z)$, so

$$u_x = \frac{A'}{f'}, \quad u_y = \frac{B'}{f'}, \quad u_z = \frac{C'}{f'}, \tag{6.137}$$

and

$$u_{xx} = \frac{A''}{f'} - \frac{f''A'^2}{f'^3}, \quad u_{yy} = \frac{B''}{f'} - \frac{f''B'^2}{f'^3}, \quad u_{zz} = \frac{C''}{f'} - \frac{f''C'^2}{f'^3}. \tag{6.138}$$

Substitution of these into (6.112) with Q given in (6.136) gives

$$\frac{A'' + B'' + C''}{f'} - \frac{\left(A'^2 + B'^2 + C'^2 \right) f''}{f'^3} = q_1 \frac{f''}{f'^3} + q_2 \left(\frac{2ff''}{f'^3} - \frac{3}{f'} \right), \tag{6.139}$$

or

$$A'' + B'' + C'' + 3q_2 = \left(A'^2 + B'^2 + C'^2 + q_1 + 2q_2(A + B + C) \right) g, \tag{6.140}$$

where $g = f''/f'^2$. If we differentiate (6.140) with respect to x, y, and z, we obtain

$$A''' = \left(2A'A'' + 2q_2A' \right) g + \left(A'^2 + B'^2 + C'^2 + q_1 + 2q_2(A + B + C) \right) \frac{g'A'}{f'}, \tag{6.141a}$$

$$B''' = \left(2B'B'' + 2q_2B' \right) g + \left(A'^2 + B'^2 + C'^2 + q_1 + 2q_2(A + B + C) \right) \frac{g'B'}{f'}, \tag{6.141b}$$

$$C''' = \left(2C'C'' + 2q_2C' \right) g + \left(A'^2 + B'^2 + C'^2 + q_1 + 2q_2(A + B + C) \right) \frac{g'C'}{f'}. \tag{6.141c}$$

Further cross differentiation of (6.141) and canceling the common term $A'B'/f'$ gives

$$2 \left(A'' + B'' + 2q_2 \right) g' + \left(A'^2 + B'^2 + C'^2 + q_1 + 2q_2(A + B + C) \right) \left(\frac{g'}{f'} \right)' = 0, \tag{6.142a}$$

$$2 \left(A'' + C'' + 2q_2 \right) g' + \left(A'^2 + B'^2 + C'^2 + q_1 + 2q_2(A + B + C) \right) \left(\frac{g'}{f'} \right)' = 0, \tag{6.142b}$$

$$2 \left(B'' + C'' + 2q_2 \right) g' + \left(A'^2 + B'^2 + C'^2 + q_1 + 2q_2(A + B + C) \right) \left(\frac{g'}{f'} \right)' = 0. \tag{6.142c}$$

Upon subtraction, (6.142) gives rise to

$$\left(A'' - B''\right) g' = 0, \quad \left(A'' - C''\right) g' = 0, \quad \left(B'' - C''\right) g' = 0. \tag{6.143}$$

This leads to

$$A'' = B'' = C'' = 2\lambda \quad \text{(constant)}, \tag{6.144}$$

as $g' = 0$ gives that $F' + F^2 = 0$ which we assume to be nonzero. If $\lambda = 0$ then A, B, and C are linear, so from (6.113) solutions are of the form

$$f(u) = ax + by + cz + d, \tag{6.145}$$

where $a - d$ are constant and it's relatively straightforward to deduce that (6.112) will admit solutions of this type, provided that Q is of the form

$$Q = -\left(a^2 + b^2 + c^2\right) \frac{f''}{f'^3}. \tag{6.146}$$

If $\lambda \neq 0$ then A, B, and C are at most quadratic, so from (6.144) and (6.113), solutions are of the form[1]

$$f(u) = \lambda \left(x^2 + y^2 + z^2\right) + k, \tag{6.147}$$

k is constant and it's relatively straightforward to deduce that (6.112) will admit solutions of this type, provided that Q is of the form

$$Q = \frac{4\lambda k f''}{f'^3} - \frac{2\lambda(2ff'' - 3f'^2)}{f'^3} \tag{6.148}$$

6.1   Exercises

1. Seek separable solutions for the following

$$\text{(i)} \quad u_t = \left(\frac{u_x}{\sqrt{u}}\right)_x \qquad\qquad u = (a(x)t + b(x))^2$$

$$\text{(ii)} \quad u_t = \left(\frac{u_x}{\sqrt{u}}\right)_x + \left(\frac{u_y}{\sqrt{u}}\right)_y \qquad u = (a(x, y)t + b(x, y))^2$$

2. Determine the form of Q such that

$$u_{xx} + u_{yy} = Q(u) \tag{6.149}$$

admits functional separable solutions of the form ([1–3])

---

[1] we have omitted the linear terms in x, y and z as the original PDE admits translation in these variables.

$$f(u) = A(t) + B(x) \tag{6.150}$$

3. Determine the form of Q such that

$$u_{tt} = u_{xx} + Q(u, u_t, u_x) \tag{6.151}$$

admits solutions of the form [6]

$$u = A(t) + B(x) \tag{6.152}$$

4. Determine the form of Q such that

$$u_t = u_{xx} + Q(u, ux) \tag{6.153}$$

admits solutions of the form

$$f(u) = A(t) + B(x) \tag{6.154}$$

5. Find conditions on $D(u)$ and $Q(u)$ (in terms of differential equations) such that the following admit functional separable solutions (see [4] and [5])

$$(i) \quad u_t = (D(u)u_x)_x + Q(u),$$
$$(ii) \quad u_{tt} = (D(u)u_x)_x + Q(u).$$

Can any of these be solved explicitly?

6. Find separable solutions to

$$u_t = (D(u)u_x)_x + \left(D(u)u_y\right)_y \tag{6.155}$$

of the form

$$f(u) = A(t) + B(x) + C(y) \tag{6.156}$$

7. The stream function formulation of the boundary layer equations is

$$\psi_y\psi_{xy} - \psi_x\psi_{yy} = \nu\psi_{yyy} + U(x) \tag{6.157}$$

Find separable solutions of the form $\psi = A(y)x + B(y)$ (Polyanin [7]).

References

1. A.M. Grundland, E. Infeld, A family of nonlinear Klein-Gordon equations. J. Math. Phys. **33**(7), 2498–2503 (1992)
2. W. Miller, L.A. Rubel, Functional separation of variables for Laplace equations in two dimensions. J. Phys. A: Math. Gen. **26**, 1901–1913 (1993)
3. R.Z. Zhdanov, Separation of variables in the nonlinear wave equation. J. Phys. A: Math. Gen. **27**, L291-297 (1994)

4. C. Qu, S. Zhang, R. Liu, Separation of variables and exact solutions to quasilinear diffusion equations with nonlinear source. Phys. D **144**, 97–123 (2000)
5. P.G. Estevez, C.Z. Qu, Separation of variables in nonlinear wave equation with a variable wave speed. Theor. Math. Phys. **133**(2), 1490–1497 (2002)
6. C.Z. Qu, W. He, J. Dou, Separation of variables and exact solutions of generalized nonlinear Klein-Gordon equations. Prog. Theor. Phys. **105**, 379–398 (2001)
7. A.D. Polyanin, Exact solutions and transformations of the equations of a stationary laminar boundary layer. Theor. F. Chem. Eng. **35**, 319–328 (2001)

Solutions

A

Chapter 1

1. $u = -\dfrac{x^2}{6t}, \quad u = \dfrac{2t}{x^2}$
2. $(a-1)(a+1) = 0, \quad 4k^2 - 2ck + 1 = 0 \quad k^2 + ck - 1 = 0$
3. $(a-1)(a\lambda + 1) = 0, \quad ck - k^2 - \lambda = 0 \quad cka + ak^2 - 2a\lambda + \lambda - 1 = 0$
4. $a = -4v, b = 2v.$
5. $a = b = c.$

Chapter 2

1. (i) $u = \dfrac{(2x+y)^2}{4}$ and $u = \dfrac{(2x-y)^2}{4}$,

1. (ii) $u = (x-1)e^{-t} + e^{-2t}$,

1. (iii) $u = \sqrt{x^2 + y^2}$ and $u = \sqrt{x^2 + (2-y)^2}$,

1. (iv) $u = e^{x+y-1}$.

2. (i) $F = F\left(\dfrac{p^2}{2} - x, q, pq - y, u - 2xp + \frac{2}{3}p^3\right)$

2. (ii) $F = F\left(\dfrac{1}{2q^2} - t, \dfrac{p}{q}, pu, 2tuq - x\right).$

3. a, b and c satisfy $a' = 6a^2, \quad b' = 6ab, \quad c' = 2ac + b^2.$

4. (i) For all k, (ii) $k = -2/3$, (iii) $k = -3/4$

5. $\dot{T} = T \ln T + kT, \quad \ddot{X} - kX - X \ln X$

6. $\dot{a} = 2ab - a, \quad \dot{b} = a^2 + b^2.$

7. $n = 2, \quad F'' + \dfrac{4FF'}{D} + \dfrac{2F^3}{D^2} - \dfrac{F^2 D'}{D^2} = 0.$

D. Arrigo, *Analytical Methods for Solving Nonlinear Partial Differential Equations*,
Synthesis Lectures on Mathematics & Statistics,
https://doi.org/10.1007/978-3-031-17069-0

Chapter 3

1. $A = -\dfrac{\phi'}{\phi}, \quad f = g + 2\,(\ln\phi)'', \quad \phi'' + (g(x) + \lambda)\,\phi = 0.$

2. $f = -\dfrac{n(n+1)}{x^2}, \quad f = \dfrac{n(n+1)}{\cosh^2 x}$

3. $F = \dfrac{1}{a}\ln v.$

Chapter 4

1. (i) $p = 3X - Y + UY, \quad q = Y - X - UY,$

1. (ii) $p = \dfrac{2(XP + 2Q)}{P}, \quad q = \dfrac{2Xe^{-Y}}{P},$

1. (iii) $p = \dfrac{e^{-X}P}{Q}, \quad q = -\dfrac{e^{-X}P}{Q^2},$

1. (iv) $p = -\dfrac{\frac{X}{Q}}{}, \quad p = U - XP.$

2. (i) $U = -\dfrac{x^2 p}{q} + c,$

2. (ii) $U = u - p^2 + c,$

2. (iii) $U = x + c,$

2. (iv) $A = c_1 p + c_2, \quad B = c_1 q + c_3,$

$U = c_1\,(u - \ln p - \ln q) + c_2\left(x + \dfrac{1}{p}\right) + c_3\left(y + \dfrac{1}{q}\right) + c_4,$

9. (i) $X = x + \dfrac{1}{q}, \quad Y = y - \dfrac{p}{q^2}, \quad U = u - \dfrac{p}{q}, \quad P = p, \quad Q = q$

9. (ii) $X = u, \quad Y = x + u, \quad U = xy, \quad P = \dfrac{x(p+1) - yq}{q}, \quad Q = \dfrac{yq - xp}{q}$

9. (iii) $X = \dfrac{x}{u}, \quad Y = \dfrac{y}{u}, \quad U = u, \quad P = \dfrac{u^2 p}{u - xp - yq}, \quad Q = \dfrac{u^2 q}{u - xp - yq}$

10. (i) $F = u - \dfrac{x}{P} - \dfrac{y}{Q}$

10. (ii) $F = \sqrt{uPQ} - x - yQ$

10. (iii) $F = 2xP + yQ - 2u - \dfrac{1}{2}P^2 - x^2$

10. (iv) No F exists

10. (v) $F = PQ - yP - xQ + u + \dfrac{1}{2}(x - y)^2$

13. $p = e^a \cos b, \quad q = e^a \sin b, \quad r = e^{-a}\sin b, \quad s = e^{-a}\cos b$

$$x = e^{-a}\,(\cos bU_a - \sin bU_b), \quad y = e^{-a}\,(\sin bU_a + \cos bU_b)$$
$$u = U_a + U, \quad v = -V_a - V$$

where U and V satisfy

$$e^{-2a} \sin 2bU_a + e^{-2a} \cos 2bU_b + V_a = 0$$
$$-e^{-2a} \cos 2bU_a + e^{-2a} \sin 2bU_b + V_b = 0$$

Chapter 5

1. (i) $F = F\left(u, \dfrac{p}{q}, \dfrac{xp}{q} + y\right)$

1. (ii) $F = F_1(x + u, p - q), \quad F = F_2\left(x + y, \dfrac{p+1}{q}\right)$

1. (iii) $F = F_1(x - q, y + p), \quad F = F_2(x + q, y - p),$

1. (iv) $F = F_1\left(y + q - \dfrac{1}{2}p^2\right), \quad F = F_2(x + yp, y + q).$

2. $P'(\rho) = \dfrac{\rho^2}{(a + b\rho)^4}$

$$F = F\left(\dfrac{\pm p + abq + b^2 pq}{bp(a + bp)}, \dfrac{ap(a + bp)x + (abpq + a^2 q \mp p^2)y + bp(a + bp)u}{abpq + a^2 q \mp p^2}\right)$$

3. $x = \dfrac{1}{X}, \quad y = \dfrac{Y}{X} - Q, \quad u = \dfrac{U}{X} - \dfrac{YQ}{X} + \dfrac{1}{2}\dfrac{Y^2}{X^2}.$

4. $x = \dfrac{1}{X}, \quad y = Q, \quad u = \dfrac{YQ - U}{X}.$

Chapter 6

1. (i) $a'' = a^2, \quad b'' = ab.$

1. (ii) $a_{xx} + a_{yy} = a^2, \quad b_{xx} + b_{yy} = ab.$

2. (a) $F'' + 2FF' = 0, \quad Q'' - FQ' - (3F' + 2F^2)Q = 0.$

2. (b) $f(u) = ax + by + c, \quad Q = -(a^2 + b^2)\dfrac{f''}{f'^3}$ for all $f(u)$, $f' \neq 0$.

2. (c) $f(u) = \lambda(x^2 + y^2) + k, \quad Q = 4k\lambda\dfrac{f''}{f'^3} - 4\lambda\left(\dfrac{ff'' - f'^2}{f'^3}\right)$ for all $f(u)$, $f' \neq 0$.

3. (a) $Q = A(u_t) + B(u_x), \quad A, B$ arbitrary.

3. (b) $Q = F(u)\left(u_t^2 - u_x^2\right) + G(u)$ where $F'' + 2FF' = 0, \quad G'' + 2F'G = 0.$

4. $Q(u, p) = \dfrac{f'' p^2 + c_1 f + C(f'p)}{f'}$ where c_1 and C is an arbitrary constant and function.

Index

© The Editor(s) (if applicable) and The Author(s), under exclusive license to Springer
Nature Switzerland AG 2022
D. Arrigo, *Analytical Methods for Solving Nonlinear Partial Differential Equations*,
Synthesis Lectures on Mathematics & Statistics.
https://doi.org/10.1007/978-3-031-17069-0